Lawrence H. Prince

The Fire Protection of Hospitals for the Insane

Lawrence H. Prince

The Fire Protection of Hospitals for the Insane

ISBN/EAN: 9783337021306

Printed in Europe, USA, Canada, Australia, Japan

Cover: Foto ©berggeist007 / pixelio.de

More available books at **www.hansebooks.com**

THE

FIRE PROTECTION

OF

HOSPITALS FOR THE INSANE

By L. H. PRINCE, M. D.

RESIDENT PHYSICIAN "BELLEVUE PLACE," BATAVIA, ILL.; FORMERLY
ASSISTANT PHYSICIAN ILLINOIS EASTERN HOSPITAL
FOR INSANE, KANKAKEE.

CHICAGO:
C. H. BLAKELY & CO.
1891.

PREFACE.

This little manual is offered with the hope that it may prove of assistance to those who are interested in the proper fire-protection of Hospitals for the Insane. While much that is contained in its pages should be of interest to architects, builders, managers, and trustees, the work is especially intended for the use of Superintendents, and other officers, attendants, and employes in general of hospitals and asylums —those who are more directly concerned in the prevention and extinguishment of fire.

The author, regretting the incompleteness of the work in many respects, trusts that it may at least suggest what is necessary to be done in order that the dangers from fire in Hospitals for the Insane and other public institutions may be greatly lessened.

While especially adapted for the use of Hospitals for the Insane, the suggestions herein offered may be easily modified to suit the requirements of all large public institutions, such as Asylums for the Deaf and Dumb, the Blind, the Idiotic and Feeble-Minded; Orphan Asylums; Sailors' and Soldiers' Homes; City Hospitals; Industrial Schools; Almshouses; Infirmaries; Reformatories; Penitentiaries, etc.

L. H. P.

BATAVIA, ILL., July 1, 1891.

CONTENTS.

CHAPTER I.

INTRODUCTORY.

CHAPTER II.

THE PREVENTION OF FIRE.

CHAPTER III.

FIRE EXTINGUISHING APPARATUS.

CHAPTER IV.

THE FIRE ALARM.

CHAPTER V.

THE FIRE BRIGADE.

CHAPTER VI.

DRILLING OF THE BRIGADE.

CHAPTER VII.

THE FIGHTING OF FIRE.

"Bitter experience has demonstrated, in innumerable instances, in this and other countries, the horrors that await hospitals for the insane whose authorities have been short-sighted—one might even say *inhumane*— enough not to provide adequate protection against the emergency of fire. Above all things, the system should be *thorough* and capable of standing the test of immediate application at a moment's notice."— *Dr. G. Alder Blumer, Utica, N. Y.*

"There is only one answer possible to the question, 'Is it desirable that hospitals for the insane be provided with a thorough system of fire-protection?' It is a self-evident proposition to which the greatest possible emphasis has been given by the series of appalling calamities from fires in insane asylums occurring year after year in all parts of the land, and I trust the time will come when a full equipment for fire-protection will be regarded as indispensable to every institution for the insane, and the necessary provision will be made to establish it in the same manner that provision is made for the ordinary expenses of each institution."— *Dr. R. S. Dewey, Kankakee, Ill.*

"There is, after all is said and done, some propriety in elaborate and long-continued, and, no doubt, irksome precautions against dangers that may never come. The daily and nightly exercise of attendants with fire-hose and extinguishers, and of patients with the fire-escapes, has been burdensome and complained of, and even ridiculed, and for several hundred days was of no immediate or apparent service, but it stood the asylum in good stead on one final day of the hundreds."— *Dr. A. E. McDonald, in Annual Report for 1883, New York City Asylum for Insane, Ward's Island.*

THE

FIRE PROTECTION OF HOSPITALS
FOR THE INSANE.

CHAPTER I.

INTRODUCTORY.

THE subject of fire protection of hospitals for the insane has never received the attention it would seem to merit. During the past few years, however, there has been a move in the right direction, which shows signs of healthful progress. The melancholy lessons of the past ought to be sufficient for all time to come. Though the burning of an asylum, accompanied by the loss of many lives, usually stimulates hospital authorities to greater efforts in providing for their institutions better protection against fire, in many instances the effect is but temporary, and after a few thousand dollars' worth of apparatus is placed in the institution and a few rules have been framed, setting forth what is to be done in case of fire, interest in the matter rapidly wanes, and soon a comfortable sense of security settles upon the members of the household, from which nothing less than another conflagration can possibly arouse them.

The danger of fire is ever present, and the fact that an institution has escaped for many years is no guaranty of safety for the future. It is too often with hospitals for the insane as it is with towns and villages: little or no thought is taken of the possibility of fire until the place is wholly or partly destroyed.

Following is a partial list of the more serious fires occurring in American institutions for the insane since the year 1850. A record of 241 lives lost by fire, the greater number during the past ten years, is in itself a sufficient argument for the better protection of our institutions.

SUMMARY OF THE MORE SERIOUS FIRES IN AMERICAN ASYLUMS.

December 3, 1850, night. Maine Insane Hospital. Origin—hot-air chamber. *Twenty-eight lives lost.*

July 14, 1857, 7 A. M. New York State Asylum, Utica. Origin—patient starting fire in large foul-air flue in attic cupola of front central building. Loss, $69,000. *Two lives lost.*

July 18, 1857. New York State Asylum, Utica, barn. Origin—fire set by patient.

——— 1860. Kentucky Western Lunatic Asylum. Loss, $200,000. *One life lost.*

——— 1862. Vermont Asylum, Brattleboro. Origin — furnace. Loss, $40,000. *Four lives lost.*

September 25, 1872, 1:15 P. M. Northern Ohio Asylum, Cleveland. Origin—supposed carelessness of mechanics; cupola of tower of Administration building. Water tanks in tower fell just as streams were turned upon fire. The falling of the tanks not only killed five persons, but rendered useless the entire water supply system by breaking pipes, pumps, etc., placed in basement. *Loss of life, five.*

——— 1872. Nebraska Hospital. *Several lives lost.*

February 13, 1877, 7 A. M. Vermont Asylum, Brattleboro. Stable. Origin—burning out of chimney. Loss, $25,000.

October 15, 1879. Missouri State Lunatic Asylum No. 1., St. Joseph. Origin—drying room. Loss, $89,000.

March 8, 1880. Kansas State Hospital, Ossawatomie. Administrative building completely destroyed. Loss, $40,000.

November 15, 1880, 7 P. M. Minnesota State Hospital, St. Peter, male wing. Origin—in basement, possibly near steam coils. "*On account of smoke, the internal means of protection could not be used.*" Loss, $100,000. *Twenty-four lives lost.*

INTRODUCTORY.

April 19, 1881, 1 A. M. Illinois Southern Hospital, Anna. Origin—in attic, from spontaneous combustion. Loss, $100,000. *One life lost.*

December 21, 1883, 10:30 A. M. New York City Asylum, Ward's Island. Origin—in attic of new east wing, from carelessness of workmen. Loss, $25,000.

January 18, 1885, midnight. Illinois Eastern Hospital, Kankakee. Origin—floor over furnace in male infirmary cottage. Loss, $25,000. *Seventeen lives lost.*

February 12, 1885, 8 P. M. Philadelphia Almshouse, Insane department. Origin—in drying room under third floor. *Twenty lives lost.*

February —, 1885. Indiana Hospital, Indianapolis. "From December, 1884, to February, 1885, we had three fires, and in each case the origin was from a steam pipe in contact with combustible matter. Cost to State, $60,000." From Superintendent's report.

June 7, 1885, 10 P. M. Virginia State Hospital, Williamsburg. Origin—right wing of center building, probably from electric light wires. Loss, $140,000. *Two lives lost.*

January 2, 1886, afternoon. Essex County Asylum, Newark, N. J. Origin—in chute in which was placed a steam coil. Spontaneous combustion. Loss, $16,000.

———— 1887. Oak Lawn Retreat (private), Jacksonville, Ill. Origin—in attic from defective hot-air flue. Loss, $20,000.

August 4, 1887. Wisconsin State Hospital, Mendota, stable.

October 12, 1887. Northern Ohio Asylum, Cleveland. Origin—in drying room. Rear buildings, including Amusement Hall, destroyed. *Eight lives lost.*

December 2, 1887. London, Ontario, Can., Hospital. Origin—in drying room. Kitchen and laundry destroyed.

March 4, 1889. Eastern Michigan Asylum, laundry and carpenter shop. Origin—from carelessness of employe.

May 6, 1890, M. St. Jean de Dieu Hospital, Longue Point, Quebec, Canada. Loss, $700,000. *Loss of life estimated at about one hundred, possibly more.*

May 7, 1890 (night). Chenango County House (Insane Department), Preston, N. Y. Origin—supposed to have originated from smouldering fire of the afternoon, a female patient having lost her life by placing a lighted pipe in her pocket. There was no night watch. *Thirteen lives lost.*

March 13, 1891, 10:15 P. M. West wing Tennessee Central Hospital, Nash-
ville. Origin—unknown. Loss, $50,000. *Eight lives lost.*

Year.	Loss of Life.
1850	28
1857	2
1860	1
1862	4
1872	12
1880	24
1881	1
1885	39
1887	8
1890	113
1891	9
Total	241

As hospitals for the insane increase in number and size,
the necessity for thorough and systematic fire protection
becomes more and more urgent. Simply providing an institution with fire-extinguishing apparatus is not adequate protection as many are inclined to believe. A thorough system of
fire protection must include measures for the prevention of fire,
a rapid method of reporting fire when it occurs, reliable appliances for extinguishing fire, and a body of well-drilled men,
ready and willing, at a moment's notice, to battle against it.
The system, to be reliable, must be sound in all its parts. A
flaw in any part will surely weaken the whole.

The fire protection of a hospital should be considered one
of the *departments* of the institution, and it ought to receive as
careful attention in regard to the details of its management as
do other departments.

It would be well, for the sake of economy, if for no other
reason, for each State to make ample provision, by legal enact-
ment, for the protection from fire of its public buildings. It
should be required that in the construction, equipment, and
management of any public institution certain *specified* precautions shall be taken looking to the prevention of fire; that each

institution shall be provided with a sufficient number of fire-escapes to allow of the ready escape, in case of fire, of all those confined in the burning building; that there shall be at the institution a sufficient amount of first-class fire-extinguishing apparatus, and that there shall be some system of drill whereby the apparatus may be quickly and intelligently used when needed.

A hospital situated within a city having an efficient fire department should not depend for protection entirely upon the city department. It should have its own private fire-fighting apparatus, and its own well-drilled brigade; it should have its own system of fire-alarm, as well as direct electric communication with the city department. There need not be as extensive an equipment nor as elaborate an organization as would be necessary for the same institution if situated in the country; nevertheless, no matter how favorably located in this respect, there ought to be a system of protection as *thoroughly* organized in every particular as though it depended entirely upon its own efforts for fighting fire. There is the same necessity for continuous effort at preventing fire as exists in any other hospital, and the same need of discovering and of fighting fire in its incipiency.

No reliance can be placed upon an indoor or stand-pipe system of fire protection. There are several objections to such a system being used, either alone or as an auxiliary to an outside system, the more important objections being: 1. The location of the fire or the density of the smoke may be such as to make the reel of hose or the stand-pipe inaccessible. 2. The progress of the fire, or the amount of smoke or heat may at any moment make it necessary for those using the hose to change their position or to retire, and to do so they must abandon their means for fighting the fire, and this not necessarily at a late stage. 3. By the burning of the hose at any point between the nozzle and its connection with the stand-pipe; by the nec-

essary abandonment of the hose while the stream is still on; or
by the breaking or tearing down of the stand-pipe by a falling
wall there would be such a serious enfeebling of the pressure
as to make almost valueless other streams. 4. In order that
all of the many wards of a large institution be supplied with
sufficient hose to cover every point, a very large amount would
be necessary. In such case a cheap grade of hose is liable to
be used, in preference to hose of good quality. Even if the
hose be of good quality it will suffer in time from non-use, as
it cannot be used for drill as frequently as would be necessary,
and if wet at any time it could not be dried as it should be.

Inside protection should consist of portable appliances
only, such as fire-pails and chemical extinguishers, to be used
for the purpose of holding in check or of extinguishing, if
possible, incipient fires, pending the arrival of the brigade.

The suggestions offered in the succeeding chapters are
based upon the following propositions:

1. Fires in hospitals for the insane, as elsewhere, are in
most instances preventable.

2. The rapid spread of fire is generally due to the faulty
construction of buildings.

3. The appliances to be depended upon for extinguishing
fire should be easily accessible, but entirely independent of the
buildings to be protected.

4. A reliable and rapid method of giving an alarm is essential to a perfect system of fire protection.

5. No system of fire protection is complete that does not
include a thoroughly organized and well-drilled fire brigade.

CHAPTER II.

THE PREVENTION OF FIRE.

I T is far easier to prevent fire than to master it. Many fires occurring in public institutions have been preventable, and, had the proper precautions been taken, the fire record would present a far different aspect than it now does. It seems strange that well-known facts in regard to danger from fire are so often entirely disregarded, or very meagerly entertained, in the matter of the construction of buildings in which are to be confined scores of helpless individuals. Mr. Edward Atkinson, President of the Boston Manufacturers' Insurance Company, says: "In the companies with which I am connected we can afford to insure cotton factories, woolen mills, paper mills and other classes of risks of like kind; but we cannot afford to insure the city warehouses in which the goods made in these factories are distributed; nor the schoolhouses in which the children are educated; nor the hospital to which the operatives are sent when ill; nor the churches where they customarily attend public worship — because they are as a rule too hazardous in their method of construction."

Surely hospitals for the insane should be as carefully protected from the dangers of fires as are mills.

The pecuniary loss is often greater at one fire than would have been the amount necessary for making the whole institution comparatively free from danger. Nor is a large expenditure of money necessary in order that a building may be made reasonably safe. At least it may be so constructed that fire will spread but slowly, thus giving those who are confined within its walls a better chance for escape, and the firemen a better opportunity for saving life and property. The expense of such construction need be but little more than the cost of building fire-traps. The prevention of fire should be very seriously considered in the construction of buildings, in the heating and lighting of the same, and in the management of the institution as a whole. Almost, if not quite, as much care should be observed in regard to the apparently *little* dangers as to the greater ones, for conflagrations are generally traceable to insignificant beginnings. Whatever is attempted with the object in view of preventing fire, in order that it be effectual, must be done in a most thorough and painstaking manner.

CONSTRUCTION OF BUILDINGS.

FAULTY CONSTRUCTION.—The rapid spread of fire in a building, from floor to floor, often directly from cellar to roof, is almost always due to faulty construction, especially of its interior, for many times, where great pains and expense are expended on solid and comparatively indestructible outside walls, the interior construction is decidedly unsafe, no attention whatever being paid to the danger of the rapid spread of fire starting within. Referring to defective and unsafe building construction, Mr. Gerhard* says:

"How are such structures usually erected and built? Their interior is of a highly inflammable character, consisting of all kinds of timber, studs, floor beams, floor boards, rafters, furring and lathing, the latter being narrow strips of wood, to which a thin skin of plaster is made to adhere. The inside of partitions, and the spaces between ceiling joists remain hollow, and form a large number of wooden flues, which constitute excellent concealed passages, not only for rats, mice, foul air and bad odors, but also for smoke and flames. By means of these thousand hidden flues a fire is spread unobserved, with astounding celerity, from the basement of a house to its attic, while it is, at the same time, most difficult to reach the flames with

* The Prevention of Fire: Wm. Paul Gerhard, C. E.

water. The foregoing description refers not only to wooden buildings, but is equally true of brick or stone houses. As usually constructed, they are no better, no safer than frame structures. It is true, stone and brick afford protection against flames from the outside, but stone or brick walls form merely the outer shell, being lined on the inside with a network of inflammable studding, furring and lathing. A fire, once started inside the building, will soon cause its rapid destruction."

If hospitals for the insane were built not more than two stories in height; if the building of them was strictly upon the principle of slow-burning construction; and if for the fast-burning wood now so generally used for interior work and roofs could be substituted some such fire-resisting wood as the California redwood,* the fire risk would be reduced to a very low point.

FIRE-PROOF CONSTRUCTION.—After mentioning some of the objections to materials generally used in so-called fire-proof construction, Mr. Gerhard says:

"Experience teaches that hard-burnt bricks, the so-called fire-bricks and terra cotta, will resist the destructive action of the flames better than iron or wood. Brick or terra-cotta fronts are better than iron or stone fronts. Iron or wood used in the construction of buildings, in the shape of columns, girders or beams, should not be left exposed, but ought to be suitably protected against the heat by a non-conducting and fire-resisting covering of either good brick-work, sound plastering, concrete, fire-clay or

* The writer is indebted to Dr. F. W. Hatch, Medical Director of the California Insane Asylum at Agnews, for the following information concerning the fire-resisting qualities of redwood: "In conversation with the chief of the San Jose fire department, he told me that it was an undoubted fact that houses built of redwood were the safest. The wood is hard to ignite, and when ignited is slow to burn; there is little flame and heat, and it is almost instantly extinguished when water is thrown upon it. It takes a minimum quantity of water to put a redwood fire out. The Chief took me to a building that had been partially consumed by fire. It was a frame composed of pine and redwood, and the difference in destruction of the two woods was marked: while the pine was burned clear through, the redwood was, in most cases, only charred on one side. * * * I am informed that after redwood has been fired and extinguished it is exceedingly difficult to re-ignite it, from the fact that the charred portion resists heat very strongly."

terra cotta. Floors may thus consist of iron I beams with flat, hollow tile arches, no part of the iron beams being exposed. Iron columns similarly protected, may be neatly finished, by plastering with smooth mortar, Keene's cement or otherwise. With all wood-work the aim should be to prevent the immediate access of air, so as to retain its strength and soundness, even during hours of exposure to heat and flames. Thus, wooden ceilings may be protected with porous terra-cotta tiles, securely fastened to the underside of joists and covered with a coat of gypsum. Timber may also be more or less thoroughly protected and rendered uninflammable by treating it with fire-proof coatings of paint, by impregnating it with chemical solutions, or by covering it in an efficient manner with plaster or some other suitable, fire-resisting substitute, such as terra cotta. No hollow wood partition walls nor any hollow wood furrings should be tolerated in such a structure. Interior walls and partitions should be made of solid blocks of concrete masonry, òr of hollow, well-burnt porous terra cotta tiles. The roof may be wire-netted on the inside, and plastered and covered on the outside with some incombustible material as a protection against sparks from chimneys or neighboring houses on fire.

"In such a fire-proof structure, a fire once started in any apartment or on any of the floors, would, in all ordinary cases, merely consume the furniture or goods contained in the room, and then die out for want of fuel, without injuring the building further than possibly scorching or burning doors, wainscoatings, baseboards, wood casings, window sashes and upper floor boards, and causing a slight additional damage to the ceilings and plastered walls by the smoke."

No hospital building of more than two stories in height should be constructed in any other manner than as nearly fire-proof as it is possible to make it. It is not necessary, however, that this expensive method be adopted for two-story buildings or cottages in order that they may be comparatively safe.

SLOW-BURNING CONSTRUCTION would probably meet all the requirements necessary for combining economy with safety in the construction of cottages and other hospital buildings, whether entirely detached or connected by corridors. The adoption of this method in the construction of hospitals would alone greatly lessen the risks from fire. "The chief principle of slow-burning construction," to quote Mr. Gerhard again, " consists *in absolutely avoiding all concealed spaces, such as hollow wood floors, furrings or partitions,* where fire can lurk or spread unobserved, and, if detected, could not be reached with streams of water."

There are various ways of carrying out this principle, with which architects are familiar. The substitution of wire netting for the usual wooden laths, and the filling in of hollow spaces with some incombustible material, such as mineral wool, slag, etc., is now often done. By filling in the hollow spaces in floors and walls the rapid spread of fire is prevented, rats and mice are kept away, and the walls and floors are deadened.

The construction of *roofs* should receive the same careful attention as do the outside walls and the interior of buildings, as there is much danger of the spread of fire from one building to another if the roof-covering is of combustible material.*

Hollow CORNICES of wood are very undesirable, as they allow of the rapid spread of fire.

EXITS.— From every ward there should be at least two separate and distinct means of egress aside from the outside fire-escapes, so that in case one way should be cut off by reason of fire or smoke, another would be available.

* "For the best construction of flat roofs in the Northern States, the roof plank should not be less than three inches thick; in the Southern States it may be two and one-half inches thick. The plank should be grooved and spliced, and covered with tin, gravel or duck. If a pitched or mansard roof is adopted — which is generally unadvisable — the plank should be not less than two inches thick, and covered outside with shingles laid over three-quarters of an inch of mortar, unless slate is used on account of close proximity of other buildings, or some other adequate reason. If slates are to be used, they should also be laid over mortar, as they are subject to almost instant destruction when exposed to a moderate degree of heat, and sparks can often pass between the cracks."— C. J. II. Woodbury, " Fire Protection of Mills."

All *exit-doors* ought to be so arranged as to swing *outward*, in order that the dangers from a jam may be avoided. This rule applies to doors leading from wards, dormitories, dining-rooms, chapels, amusement halls and other places where a number of patients are congregated; it also applies to doors leading to fire-escapes. Very many lives have been lost in the past, during fires and panics in churches and theaters, as is well known, owing principally to the fact that the exit-doors were made to swing inward.

Staircases and *Hallways* require and should receive special attention in the matter of fire-proofing. The *stairs* should be made of the very best material, and whatever is used the under-side ought to be made as nearly fire-proof as possible. They should be of ample width, with no sharp turns, a landing being placed where a turn is necessary. Where the stairs fill up a hallway, hand-rails should be provided.

FIRE-ESCAPES.— The ordinary vertical or nearly vertical iron ladders, used so much as fire-escapes on large buildings in cities, should never be used for this purpose on any hospital building. Of course they are better than no escape at all, but very little better. Under the most favorable conditions it would be no easy task for an ordinary man in good health to descend from a second or third story upon one of these. How would it be, on a cold winter night, for frightened, insane men and women, with bare hands, to grasp the frosty iron rungs and attempt a descent? The only truly practicable fire-escape is one that will answer the purpose of ordinary stairs, and be as easy of access. There should be a platform or balcony, and a pair of stairs, the latter running parallel with the side of the building, protected by a strong railing, the whole being made of iron — as much as possible of gas-pipe, [Fig. 1.] Wooden treads are to be preferred to iron, however, owing to the greater liability of the latter to become coated with ice. It is important that there be an easy means of exit from the building to these escapes. The best plan, and the only feasible one, is to have a door opening on to the escape from each ward or domitory above the first floor, *the door opening outwards, and controlled by the common ward key.* The size, number, and arrangement of

FIG. 1.— Iron fire-escape and balcony fire-escape.

escapes will necessarily depend upon the size and shape of the buildings upon which they are placed. In any case, this rule should be adopted,— that *fire-escapes should be of such construction, and placed in such positions, as to allow of the ready escape of any number of persons from a building, when from any reason the stairways are rendered inaccessible.*

Fire-escapes of this kind answer a useful purpose, also, in furnishing firemen an easier, safer, and quicker means of reaching with their hose a second or third story than do ladders or inside stairways.

Properly constructed *balconies* of cottages may be utilized as fire-escapes, dormitories opening directly upon them, a pair of stairs leading from the second to the first floor of the balcony. [Fig. 1.] The stairs may be hinged and weighted, so as to be drawn up during the day, or be made stationary.

FIRE-DOORS.— It is doubtful whether fire-doors, except where used in connection with a fire-wall, or in fire-proof connecting corridors, would to any appreciable extent prevent the spread of fire. The tinned wooden doors are greatly to be preferred to those of iron.

In writing upon the subject of fire-doors for mills, Mr. Woodbury * says:

"A dear-bought experience has shown iron to be absolutely unfit for such purposes. The heat of a slight fire will cause an iron door to sink by its own weight, while the writer has seen double 'air-space' doors, made of thin corrugated iron, shrivel like the withered leaves of autumn, when the fire was insufficient to destroy the paint on them. The most efficient fire door is constructed of two thicknesses of tongued and grooved inch boards laid diagonally across each other, and nailed with wrought iron nails and driven flush, clinched on the other side. This door is then covered on sides and edges with sheets of tin locked together like a tin roof. The lintel should be tinned and securely fastened to the masonry. If a swinging door, the hinges should be securely bolted in, and not merely fastened with screws."

ATTICS AND BASEMENTS.— Fires having their origin in attics and basements usually make considerable headway before being discovered, owing to the fact that these places are much less frequented than are the main floors. No rubbish or dirt of any kind should be allowed to accumulate in basements, nor should attics be made receptacles for lumber, old furniture, bedding, or anything else that would make fuel. These places should be frequently inspected, and always be made inaccessible to those having no business in them. As many fires have occurred through the carelessness of plumbers and tinners working in attics, basements, and other out-of-the-way places, it would be well, as a matter of precaution, to follow up these folks with pails of water or fire extinguishers.

* "The Fire Protection of Mills," by C. J. H. Woodbury.

LAUNDRY AND DRYING-ROOMS.— Owing to the danger of fire from laundry apparatus, especially that used in the ironing and drying rooms, this department should be located in a separate building from the hospital proper. Great care should be taken that clothing, rags, and other combustible articles do not come in contact with nor accumulate about the steam pipes, stoves, ranges, or other apparatus used for heating and drying purposes. Automatic sprinklers may with advantage be placed in the drying and other laundry rooms, other precautions not being neglected, however. Special precautions are necessary in these places for the protection of all exposed woodwork, and inspection should be frequent and thorough.

ELEVATOR AND DUMB-WAITER SHAFTS are often responsible for the rapid spread of fire. Through them the flames will pass quickly from basement to attic, converting in a few moments an otherwise feeble, easily managed blaze, into a fire difficult or impossible to control. There should be nothing, throughout the whole extent of these places, for the flames to feed upon, the sides being constructed either of some uninflammable material, as brick or stone, or else lined with tin or galvanized iron. All of the doors ought to be tin-lined and made to fit snugly. Except when in use the openings into the shafts should be kept closed.

FOUL-AIR FLUES AND DUST CHUTES are a source of danger in the same way as are dumb-waiter and elevator shafts. The fewer the number of openings there are between one floor and another the better. Where they are indispensable the danger from them should be borne in mind, and proper precautions taken to prevent accidents.

HEATING OF BUILDINGS.

Everything pertaining to the warming of hospitals ought to be given very careful attention. In this matter great care is necessary that none of the minor details be neglected. It is in the neglect of the apparently insignificant matters in connection with the putting in and management of heating apparatus that often leads to disaster.

FURNACE-HEATING.— The *room* or *compartment* in which a furnace is placed should be made thoroughly fire-proof — the floor, walls, and ceiling — whether the furnace-dome and sides are insulated or not.

The *furnace* should be of such size that it need never be necessary to crowd it in order that the building may be warmed in severe weather. It should be constructed in the very best manner, and kept in good order and repair. The furnace-dome and sides should be incased in some heat-resisting material, and placed at a *safe* distance from combustible substances, the distance, of course, depending upon the amount of heat the furnace is capable of giving off, and the inflammability of the surroundings. Never, under any circumstances, ought an uncovered furnace be placed beneath an unprotected, combustible ceiling. The walls of the furnace should be of brick, of ample thickness, and hollow, admitting of an air-space of about four inches.

Smoke-Pipes should never pass through a floor unprotected. Where entering a flue, the pipe should be at least twelve inches from floors, ceilings, or partitions, unless a metal shield is provided, in which case six inches space is sufficient. Where the smoke-pipe passes through partitions the same precautions are necessary as for heat conductors. The ceiling above the pipe, unless it be fire-proof or at least eighteen inches away from the pipe, should be protected by a metal shield, not fastened to the ceiling, but *separated from it by an air space*. Tin shields fastened against wood-work are not as safe as generally considered. The wood often chars, tin being a good heat conductor.

Cold-air Chambers may become a source of danger from fire, in the event of a reversal of the draft, if not properly constructed and protected. They should be made entirely of non-combustible material, be kept free from inflammable substances, and the mouth or opening of the box covered with wire netting, to prevent the blowing in of paper, rags, etc.

Heat-Conductors ought to be made double, one larger metal cylinder enclosing a smaller one, a space of at least one inch separating the two. Where the conductor passes beneath or close to wood-work, there should be placed between them a shield of tin or galvanized iron, an air-space being left on either side. When passing through wooden partitions a collar of tin, made double, and ventilated, should encircle it, or it may be enclosed in masonry or brick-work. Especial care must be taken in the matter of fire-proofing where conductors pass through closets and clothes rooms, where there is but little change of air, and where inflammable articles are so liable to be thrown about them. Upright pipes in walls should be enclosed in brick-work or other fire-resisting material. No hot-air flues or conductors should pass between a combustible floor and ceiling.

Hot-Air Registers should be placed in side walls, and preferably near the ceiling, out of the reach of mischievous patients. The wood-work about them should be thoroughly protected by soapstone borders set in plaster-of-Paris or gauged mortar. A wire netting fastened over them, on the inner side, will prevent the access of dirt of any kind. Fans or valves should either be entirely discarded, or else made accessible to responsible persons only. Great danger arises from the confining of heat in hot-air pipes or conductors by the closing of register valves. The principal register of a furnace should never be closed.

STEAM-HEATING.— The same precautions suggested in reference to *furnaces and furnace-rooms* may be applied, wherever practicable, to the furnaces and furnace-rooms where steam-heating is used. It is better to have the boiler located in a separate building.

There is a popular though erroneous belief that hot steam pipes and radiators are devoid of danger when in contact with or in close proximity to wood or other combustible material. Many fires have originated in this way, and the danger is not by any means slight, especially when the high-pressure system is used.

Steam Pipes become a source of danger when they pass through floors and partitions where no provision has been made to protect the combustible surroundings. The lathing, when in contact with the pipes, gradually becomes charred through friction of the hot pipes, thus becoming more and more inflammable. If the pipe is unpainted, it sooner or later rusts, and the action of rust upon carbonized lath, under certain favorable conditions, produces combustion. Examples of such accidents are not wanting. A coating of paint will prevent pipes from rusting, and this should be done where pipes pass through floors or ceilings, or in close proximity to wood work. Pipes passing through floors or partitions, whether painted or not, ought to be separated from its surroundings by a metal thimble, so arranged as to allow of an air-space between it and the pipe. The material used for insulating steam pipes should be practically indestructible, at least by heat. Prof. Chas. B. Gibson, in " Hazard of Steam Pipes," says:

"All organic matter, such as hair, felt, shoddy and paper, becomes more or less charred by constant contact with hot steam pipes for a long time, even though the temperature be but little above the boiling point of water; and by steam of 300° F. and above, so thoroughly scorched after a time as to become very fragile and to crumble away rapidly. It is noticeable that the dust formed from this charred material is very combustible, and will flash like gunpowder when thrown into a fire; hence, it is evident that if it becomes once ignited, rapid combustion will ensue."

In *indirect radiation* in steam heating the same careful attention should be given to hot-air flues and registers as in furnace heating. The surroundings of steam-traps, indirect radiators, etc., ought to be kept free from dirt, and all wood-work be properly protected.

Radiators and *coils* should be placed a safe distance from the wall or ceiling, and dirt, rags, chips and clothing never allowed to collect about them, as it does not take a great while for a sufficient amount of heat to accumulate between the coils when confined to set fire to any combustible articles in contact with them. Even though there be but a small amount of heat given off, sufficient may accumulate, under favorable conditions, to prove a source of danger.

STOVES.— Where stoves are used, the floor should be thoroughly protected, not by a sheet of zinc alone, nor simply by a layer of brick. The brick should be separated from the flooring by a layer of cement, and over the brick may be placed the zinc. This method would insure the greatest amount of safety.

The *stove* should be free from cracks. If it is placed near wood-work of any kind, which should be avoided if possible, the exposed objects should be protected by the use of tin or galvanized iron shields. Wood must not be placed near to or under stoves, nor leaned up against them. If possible, the stove should be made inaccessible to patients.

The *stovepipe* should fit snugly at all its joints. Where the pipe is long, the joints should be riveted, and the pipe firmly supported by wires. When it is necessary that stove pipes should pass through floor or partitions, the surrounding woodwork must be protected in the same manner as are heat conductors. The pipe near floors, or where passing through closets or bath-rooms, should be surrounded by tin or galvanized iron, an air-space of two inches intervening. The tops of pipes should be frequently dusted. The stovepipe should not enter the chimney in an unused room or in a closet.

HOT-AIR FLUES.—It has been estimated that 25 per cent. of all fires occurring in the United States are caused by defective flues, and that the annual loss thereby amounts to

$5,000,000. The defective flue is the result not so much of ignorance as of carelessness on the part of the builder. In the larger cities provision is made in the building laws concerning the construction of chimneys and flues. Hospital authorities should make themselves familiar with what is necessary in the construction of flues and chimneys, and should see to it that this source of danger is eliminated. All unused flue-holes should be closed with metal stoppers.

Ashes should never be thrown into barrels or boxes; metal receptacles only should be used.

LIGHTING OF BUILDINGS.

Under this heading will be considered the various methods in use for the lighting of hospital buildings, with some of the dangers connected therewith, and some of the ways in which these dangers may be averted. Whatever system of illumination be used, whether it be by kerosene oil, gasolene, gas, or electric light, constant watchfulness and care is necessary in order that it may be kept in as safe a condition as possible. It is poor policy to consider *any* of the methods now in use for lighting purposes as absolutely trustworthy, for none of them are. Electricity is fast taking the place of other systems of lighting, and is without doubt the safest and best of any yet employed. But few institutions are using kerosene oil, some still cling to gasolene vapor, while the majority, probably, are burning coal-gas.

KEROSENE-OIL LIGHTING.—Where kerosene oil is used for lighting purposes, the oil should be of the very best quality. If kept in large quantities it should be stored in a fire-proof room or in a separate building. It should never be necessary to enter the oil-room with a light of any kind.

LAMPS.—Glass lamps are dangerous, and ought never to be used. Those made of metal alone are objectionable on account of their liability to leak. Glass lamps in metal cases are free from the objections of either the glass or metal lamps, and are considered quite safe. A leaky lamp, or one with a loose handle or base, should never be lighted. Lamps should be cleaned and filled early in the day — *never* after dark. Filling a lighted lamp is a most dangerous thing, and ought never be permitted. Hand-lamps should never be left with patients or set about where patients can get them. It is far better to have safe hanging-lamps, placed out of reach in each hall or ward. Bracket-lamps, if used, should not be placed in proximity to curtains or draperies, and ought to be so arranged as not to swing against the wall or close to inflammable articles.

LANTERNS.— The same precautions as to the filling, trimming and cleaning of lamps, as well as to the quality of oil used in them, applies to lanterns. It is usually necessary to "pick up" the wick of a lantern once or oftener during the night, when used by night-watches especially. This should never be done on a table containing inflammable articles of any kind. A small particle of the burning wick may be flirted into the folds of the table-spread, or into a basket of paper, fancy-work, etc., and in this way start a fire. This accident has actually happened. Broken or insecure lanterns have been the cause of many fires. The lamp of the lantern should be firmly fixed, not simply held in place by a catch or spring.

GASOLENE-GAS LIGHTING.—The greatest care is necessary, in lighting a building with gasolene-gas, that the machinery is of the very best make, and that the workmanship on the pipes, etc., is perfect. The fluid itself should never be kept in the building, but always in a separate fire-proof structure or underground vault, a safe distance away. The pipes should all incline towards the machine. The vapor of gasolene is highly inflammable, and a most dangerous thing to have about in large quantities. A very careful and thorough inspection of pipes and other fixtures should be frequently made, so that they may be kept in as perfect a condition as possible.

Probably the safest way to do with a gasolene-gas plant is to get rid of it as soon as possible, substituting for it something safer and better. The storing of gasolene in the same building in which the insane live, as is sometimes done, cannot be too strongly condemned. To lock up scores of human beings in a building with a great tank of gasolene should not be countenanced for a moment.

COAL-GAS LIGHTING.—As in the case of gasolene, the piping and fixtures should always be in good repair, the joints so securely made as to allow of no leakage of gas. Frequent inspection is necessary, such inspection being done only by daylight. To look for a gas leak, whether from pipes or metre, with a candle, lamp or lantern, is the height of folly. Many fires have been caused by this carelessness, and not a few lives lost.

In halls or wards where there are disturbed or violent patients the jets should be placed entirely out of reach. Where lights are kept burning through the night, they should be far enough away from transoms to be out of reach of ropes made of twisted straw or bedding.

Metal or porcelain shields should be hung above fixtures wherever the distance between the jet and ceiling, wood-work or other ignitible substance is less than three feet. Brackets on side walls should be stationary, or else fitted with guard rings of large diameter. The wall may be further protected by a plate of metal. The fixed lights should be placed so that in opening doors they will not come near to the gas flame. Especial care must be taken in clothes closets, wardrobes, store-rooms, bath-rooms, basement rooms generally, narrow hallways and all other places where clothing and other inflammable articles are liable to come in contact with a flame. It is a very good plan to place securely about jets, in these places, large, strong wire nettings or cages.

The hanging or placing of holiday trimmings, as evergreens,

cotton batting, streamers, colored papers and the like, about gas or gas-lamp fixtures, should never be permitted.

MATCHES.—Stringent rules should be adopted in every hospital regarding the care and use of matches. They should always be kept under lock and key, and every precaution taken to keep them out of the hands of patients. It is also important to keep matches safe from rats and mice. "Safety" matches, so-called because they can only be ignited upon certain prepared surfaces of boxes, are now very extensively used in hospitals, and generally prove quite satisfactory. Care should be taken that the boxes, when empty, are *destroyed*, not thrown away, to be found later by mischievously-inclined patients. The best plan would be to require that empty boxes be returned to the store-keeper before a new box is issued. A hint or two in regard to these "safety" matches may not come amiss. Owing to the explosive character of the material of which they are made, the striking of one upon the box causes numerous sparks to fly about. A spark coming in contact with the exposed heads of the matches in the box will set fire to them, and considerable of a flash will result. For this reason the box should always be *closed* before the match is struck. When the boxes are placed upon the box-holders, fastened to the wall near lamp-brackets or gas-fixtures, it should be arranged so that the heads of the matches are at the lower end of the box.

Burned matches should not be immediately thrown into waste-paper baskets or dust-boxes, as they will sometimes retain enough heat to ignite inflammable substances.

CANDLES should be used only when absolutely necessary, and then by those who can be trusted.

ELECTRIC LIGHTING.— While the incandescent electric light which is now being adopted by so many institutions is by far the safest of all known means of artificial illumination, yet it may prove a source of danger from fire if the work of putting in wires, dynamos, lamps, etc., is not properly and skillfully done. Says Mr. Gerhard: *

* * * "The whole work should always be done under the immediate superintendence of skilled and experienced electricians. It should also, from time to time, be carefully inspected and tested. The current generating machine, or dynamo, should be fitted up in some dry place, kept scrupulously clean and free from all chips, waste or inflammable dust. Electric light wires should be run so as to be readily accessible, easily inspected, and repaired without causing undue trouble. All joints in the wires should be made perfect, and all wires in the interior of an institution ought to be thoroughly insulated, and shielded against moisture by non-conducting coverings, and protected from injury or contact with telegraph or telephone wires, and their position should be exactly laid down on the floor plans of the building. Wherever electricity is conducted into a building from outside sources, some approved automatic cut-off should be arranged near the entrance of the building, by means of which the circuit may be broken in case the generating current should become excessive.

"Wherever arc lights are employed to light staircase and entrance halls, care should be taken that the dropping carbons may not cause a fire. It is best to enclose arc lights with glass globes and provide a wire netting under the globes to keep the pieces of glass from falling in case of fracture."

The following is extracted from a paper, read before the National Association of Fire Engineers, by Mr. Wm. Brophy, Inspector of the New England Insurance Exchange. First, in regard to the insulation of electric light wires:

"The insulation most universally used for outside wires is the much talked of "underwriters" or painted cotton insulation. It is a very good insulation if kept dry, but when wet becomes an excellent conductor. I would advise you to treat any wire carrying high potential currents with the greatest consideration when it is covered with this material, and to give it all the territory you can spare when it is water-soaked from any cause.

"There are other grades slightly better than this, but they are only attempts to produce something cheaper than the higher grade."

In reference to the danger of fire from electric light wires Mr. Brophy says:

* "The Prevention of Fire," by Wm. C. Gerhard.

* * * "Where an electric lighting plant is properly installed, whether the current be furnished from a central station or a dynamo on the premises, the danger of fire occurring from the same is reduced to a minimum. The causes which lead to fires from the arc-light system are very few indeed. The amount of current being constant, and far below the safe carrying capacity of the wires, danger of over-heating the same is avoided. Imperfect joints, loose connections, or any other obstructions of this nature to the passage of the current may and sometimes do cause fires, the electromotive force being sufficient to overcome such obstructions, and, in doing so, setting fire to any inflammable material intervening. The arc lamp itself has been the cause of the greatest number of fires — pardon me, I should say the incompetent, negligent attendant, by leaving bottoms intended to be closed, open, or by not removing broken globes when discovered. Dangers of fire from the direct, low-tension system of incandescent lighting are or may be more numerous, providing they are permitted to exist. The pressure of electromotive force is nearly constant and is far below that point considered dangerous to life. The amount of current varies with the amount of light required. In large stations the currents from several large dynamos are sent into common mains or feeders. The current is sufficient in quantity to heat to a dangerous degree or even to melt the small house mains down if for any cause the resistance of the circuit should fall greatly by reason of ground connections or short circuits. To prevent this, metals of a low-fusing point are inserted, which melt long before the temperature of the wire reaches a dangerous point. Poor insulation and lack of ample separation is another source of danger. In wet or excessively damp places, wire should rest on nothing but insulating supports, no matter how good the insulating covering may be. They should not rest on wood or be fastened under wooden cleats.

"A safe rule to apply to all wires concealed between floors and ceilings and behind partitions, is to use the same care as would be necessary were they not covered with any insulation at all. The general practice in New England is to use the very best insulation for this purpose, and by ample distance between the wires themselves, and gas and water pipes, to avoid danger from fire. The transformer or silent dynamo, used in the system of that name, has been excluded from the interior of buildings in New England. The reason for this is that the insulation sometimes burns slowly, and a considerable amount of smoke results."

KEROSENE OIL AND GAS STOVES are often used at night for heating milk, etc. The stove should stand in a large shallow pan, when lighted, and entirely removed from the vicinity of combustible material. The rubber tubing should be sound and fit tightly at either end. In turning off the gas from a gas-stove, the supply should be cut off first from the gas fixture.

OILED RAGS, WASTE, ETC.—The disposition of oiled rags, cotton waste, etc., should receive very careful attention. The spontaneous combustion of oiled rags has been the cause of a very large number of fires, and probably the origin of many of the "cause unknown" fires has been due to this accident. Cloth, cotton waste, or dust, containing oil in small quantity, if placed under conditions favorable to the accumulation of heat generated within it, is liable to ignite spontaneously. It is therefore important that these substances should not be allowed to collect in closets or boxes. Dust and dirt boxes should be emptied daily. Oiled rags should either be hung up so that the air may circulate freely about them, or else spread out — never thrown in a heap into a box or closet, as is often done. Instead of being thrown away when useless, they should be burned, as rats will often carry them into walls and floors.

COAL.—The spontaneous ignition of bituminous coal, when damp and stored in poorly-ventilated coal-houses or cellars, is not an uncommon occurrence, and measures should be taken to prevent it. Coal houses should be constructed so as to provide for the proper ventilation of the coal. The coal should also be kept as dry as possible. In placing coal in furnace rooms it should not be piled against furnace walls. If the lower stratum of coal shows signs of heating it must be spread out and exposed to the action of the air.

SMOKING.—Fires are sometimes caused by the careless dis-posal of ashes from tobacco-pipes, and of recently-smoked pipes, cigars, and cigarettes. Patients sometimes place in their pockets the pipes they have been smoking, without first empty-ing out the hot ashes. Smoking about a hospital, except in certain places, should be prohibited.

Cuspidors should be of metal, and filled with sand, cinders or gravel, rather than sawdust.

WORK-SHOPS.—In the carpenter, machine, paint, and other shops, as well as in and about all of the stable and farm buildings, the matter of the prevention of fire should receive very close attention. All of these places should be frequently inspected, and the fire regulations rigidly enforced.

NIGHT WATCHES.—No institution, large or small, should be without an efficient night-watch service. This is one of the most important of all the means to be taken for the prevention of fire, and ought to receive special attention. The duties of night watches in regard to the prevention of fire should be plainly marked out. They should keep a sharp lookout for those things which, if neglected, might be the means of causing fire. Attics, basements, clothes closets, broom closets, laundry and drying rooms, kitchens, furnace-rooms, work-shops, and in fact all portions of an institution, should be visited and carefully inspected at least once during the night. The night watch should see especially that radiators, steam pipes and registers are free from inflammable articles, that furnaces are not overheated, and that closets and chutes contain no oily rags or waste.

INSTRUCTION TO EMPLOYES.—There should be embodied in the rules of every institution explicit instructions regarding that which pertains to the prevention of fire, and these should be as rigidly enforced as any other of the hospital rules. It should be the duty of the Chief Marshal or one of his assistants to explain occasionally to employes the many ways in which fires may occur, and to point out the remedy. It would be well to make this a portion of the training-school curriculum.

CHAPTER III.

FIRE-EXTINGUISHING APPARATUS.

WHETHER a hospital for the insane be of fast- or slow-burning, or so-called fire-proof construction, it should be provided with ample facilities for extinguishing fire. Situated, as institutions of this kind usually are, some distance from cities or towns, they necessarily depend, either entirely or to a great extent, upon their own resources in case of fire. Even though located near or within a city having an organized fire department, there is always more or less delay in the arrival of the firemen. To have the proper fire-extinguishing facilities always at hand, with well-drilled men located in the institution, would seem to be in all cases the very best plan when the value and importance of time are taken into consideration. Surely no institution whose location is remote from a competent fire brigade should be without its own means for fighting fire.

Fire-extinguishing apparatus, to be reliable, must be of the very best material and workmanship; it should always be in perfect working order, and at all times properly cared for. To suppose that fire apparatus of any kind will never require care, is a serious though quite common mistake — a mistake that often leads to disastrous results, as well as to condemnation of a really superior thing. The apparatus must be of such construction as to allow of considerable service, that firemen may be frequently and properly drilled in its use. Apparatus that cannot stand the wear and tear of frequent use for practice or drill is not fit to be placed in any institution, and should never be depended upon. Good apparatus is all the better for being used, and with proper care will last nearly, if not quite, as long as it would were it never handled. A good piece of machinery will rust out sooner than it will wear out.

To furnish a hospital with a cheap grade of fire-extinguishing apparatus, known to be unreliable, and simply exhibited for the purpose of warding off public opinion, is not far removed from criminality.

The kind and amount of apparatus needed in any institution will depend largely upon the size, construction, and the number and the distribution of the buildings to be protected, as well as upon the proximity of the hospital to a city with adequate facilities for fighting fire.

There are many kinds of fire-extinguishing apparatus, but it is the aim to call attention particularly to that with which the writer is most familiar, and which he considers best adapted to the purpose of fire-protection in hospitals for the insane.

Many things only lightly touched upon under the following headings, especially as to the *use* and *care* of apparatus, will be more thoroughly dealt with in the succeeding chapters.

WATER-SUPPLY.

From whatever source water for fire purposes is obtained, the supply should be *constant and practically inexhaustible*.

Water tanks should never be placed in attics nor upon the tops of buildings. They are not only unreliable for the purposes of fire protection, but more than this, they are, in case of fire in the building in which they are placed, a great source of danger from their liability, on account of their great weight, to fall.

THE FIRE PUMPS.

should be located at a safe distance from any of the main buildings, and they should be kept in such condition as to be in readiness, at any time of the day or night, to furnish water at fire pressure in from two to four minutes after the sounding of the alarm.

WATER MAINS.

The mains for the hydrant system should form a loop about the hospital plant, and ought not to be less than six inches in diameter, except where running to but one hydrant, in which case four-inch mains could be used. The mains should be supplied with straight-way stop-valves, so placed as to allow of the tapping of the main for pipe connections, or for the removal of a hydrant, without throwing out of service any but a small portion of the system.

The running of mains under buildings should be avoided; if this is not possible they should be placed deep enough so as to be safe from injury by a falling wall. Where large branches run into a building they should be provided with stop-valves, placed outside of the building at a safe distance from its walls.* The breaking of a main or large branch by a falling wall would render useless or at least seriously cripple the rest of the system.

FIRE HYDRANTS.

That hydrant which offers to the flow of water through it the least amount of resistance, is obviously the best hydrant to use, for the less frictional loss there is at the hydrant the less the amount of pressure there will be required at the pumps to give a good fire-stream. The principal cause of the loss of pressure in hydrants has been the obstruction of the water-way by the valve and valve-rod mechanism, which, when the hydrant is opened, occupies the center of the hydrant-barrel. Improvements have been made in this respect

* J. B. Freeman, H. E.

from time to time, as the change from the "globe" to the "gate" valve. With the improvements there has been a lessening of the frictional resistance. The most recent change for the better has been in removing the rod and valve from the water-way entirely, when the hydrant is open, as shown in Fig. 2, leaving a clear, uniform water-way, the same size as the main to which it is attached.

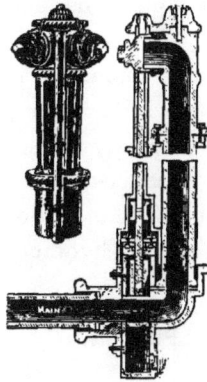

Fig. 2.—The "Beaumont" hydrant, showing hydrant open.

In planting hydrants, they should be so placed as to allow of the concentration, upon any point, of two or more streams from lines of hose no more than 250 feet in length. The fewer the number of hydrants the more hose will it be necessary to carry on the carts. For several reasons extra hydrants are to be preferred to extra hose: hydrants and hydrant mains are cheaper than hose; the larger amount of hose requires heavier carts, and this extra weight would require more men to move it; a long line of hose is more difficult of management than a short line, and more time is consumed in laying it; the loss of pressure by friction increases with the increase in the length of the hose.

The space about the base of hydrants should be filled with loose stones and covered with coarse manure, so as to form a reservoir into which the hydrants can empty.

Nothing smaller than a four-inch hydrant should be used. Those with but one nozzle are to be preferred, but where single-nozzle hydrants cannot be placed sufficiently near to each other, then two- or three-nozzle hydrants may be used to advantage.

Hydrants should be placed far enough from buildings to escape being injured by falling walls. They should be frequently tested, winter and summer. The coupling should be occasionally treated with a little mineral oil.

It ought not to be necessary to use forcible effort to close a hydrant and make it water-tight. If this is necessary it shows that the hydrant is not in good order, and unless the trouble is soon remedied the hydrant is in danger of being seriously damaged.

FIRE PAILS.

There is no doubt that many incipient fires may be extinguished by the use, at the proper moment, of a pail of water; and where several pails are handy a larger blaze may be controlled or held in check until the arrival of more efficient apparatus.*

There should be provided for each hall or ward from five to ten fire pails. By placing them in a cabinet, the lock of which is the same as those upon the ward doors, they may be always easily accessible to attendants, and at the same time be out of the reach of patients.

There are several kinds of fire pails, some of which are quite expensive. For hospital use something simple and dura-

---

* It is a matter of record that of the losses in mills *paid for* by the insurance companies, twice as many fires are put out by pails as by any other means.—Woodbury, "The Fire Protection of Mills."

ble is the best. The pail should be light, strong, and of good size. Those made of indurated fibre are superior to the common wooden pail, or to those made of paper or rubber. They are seamless, hoopless, and are said to be proof against the action of water or of the atmosphere. On each pail should be stenciled, in plain letters, "*Fire*," or "*For Fire Only*," as this will guard in a measure against their being used for other purposes. They should be kept nearly full of water, and the water should be changed at regular intervals, so that it may be kept fresh. It is better *not* to place covers upon the pails, as is often recommended, to guard against evaporation, for the reason that they can be inspected to better advantage if uncovered; and if the water is changed at regular intervals evaporation will cut no figure.

The pails made with rounded bottoms, which when filled with water, must be either hung up or set in openings in shelves or racks, while not likely to be used for other than fire purposes, are objectionable for the reason that they cannot be set down, as would in many cases be necessary, especially if a hand pump were used.

A small hand force pump, with rubber tubing and nozzle attached, would be a valuable addition to each set of pails.

BATH TUBS.

The bath tub is a useful auxiliary to the fire pail, especially where the use of the pails is alone to be depended upon in case of fire, and in those portions of buildings the least accessible to firemen. It ought to be partially filled with water every night, and the key of the faucet left in place, or quite handy. The water could be turned on by the first one coming to refill a fire pail, and thus a constant supply could be had.

HAND GRENADES.

in the shape of bottles and tubes, generally made of glass, containing a pint or more of some chemical fluid, are familiar to almost everyone, as they are to be found in a great many public buildings, manufactories, railway carriages, etc. There is no doubt that hand grenades have been the means of extinguishing incipient fires, but unfortunately the great majority of fires start in such inaccessible places as to render this means of extinguishment utterly useless. Everything considered, hand grenades are not nearly as efficient as are pails of water for fighting incipient fires. The objections to their use in hospitals so outweigh their good qualities that they cannot be recommended.

PORTABLE CHEMICAL FIRE EXTINGUISHERS.

The Graham method of extinguishing fire, by the use of a commingled stream of carbonic acid gas and water, is without doubt, next to the use of large quantities of water, the most efficient and reliable known, and the portable hand fire extinguishers from which this stream is thrown should be an essential part of the fire-extinguishing apparatus outfit of every hospital. A very large percentage of all fires occurring in institutions for the insane could be extinguished by this means, provided, of course, the fire was not allowed to gain too much headway; and even fires of no mean proportions may be extinguished or at least controlled by one or more of the extinguishers, properly used. The principal points in favor of this method of extinguishing small fires are:--- (1.) The comparative ease with which a fire, inaccessible to all other means used, may be reached; (2) the manner in which a fire is extinguished without the use of large quantities of water; (3) the accuracy with which a stream may be directed; (4) the ease with which the extinguisher is recharged and made ready for use.

These extinguishers, known as the "Babcock" and "Champion," are made of copper, in three sizes, with a capacity of six, three, and one and one-half gallons respectively.

The "Babcock."—A solution of bi-carbonate of sodium in water nearly fills the tank. A sealed bottle of sulphuric acid ("A," Fig. 3), supported in a cup ("B," Fig. 3), and held in place by a spring cap which fits over the neck of the bottle (Fig. 4), is suspended in the alkaline solution, and is firmly fixed by its connection with the top of the extinguisher, the top being tightly screwed into place. Turning the wheel on top to the left, pressure is brought to bear upon the weak portion of

Fig. 3.   Fig. 4.

the bottle noticed in figure 4, and it is broken. Immediately upon the mixing of the two fluids a large quantity of carbonic acid gas is formed, which, by its pressure, 100 pounds to the square inch, forces from the tank, through the hose attached to it, all of the contained water, together with the gas itself, a non-supporter of combustion, and sulphate of sodium, which latter is precipitated upon objects with which the fluid comes in contact. This precipitate, a white powder, is harmless, and easily removed. There is, however, liable to be an excess of the acid at the very beginning of the stream, which might prove more or less harmful to clothing or furniture. This does not always occur, but is to be borne in mind. If the stream is directed only upon the fire no damage can be done. The stream from a six-gallon extinguisher may be thrown a distance of about 35 feet, and can be controlled by a stop-cock at the nozzle. About four minutes are required for the extinguisher to empty itself. The stream may be made larger in diameter, and the contents of the tank made to escape more rapidly, by increasing the bore of the nozzle.

THE "CHAMPION."—The difference between the "Babcock" and the "Champion" is that in the "Champion" the acid is placed in an open cup in the extinguisher, and it is only necessary that the tank be inverted to cause the cup to be emptied. On this account the "Babcock" is greatly to be preferred, as the extinguisher is very likely to become inverted long before the fire is reached.

The larger size of the "Babcock" extinguishers should be placed about the institution where most needed. They should be conveniently located, easily accessible to attendants and employes, but out of the reach of patients. There should be two placed in the Amusement Hall, one or two in the administrative department, a few in the workshops, laundry rooms, etc., and four on the ladder truck, if there be one. A few of the medium and small-sized extinguishers should be placed about the wards for female patients.

In purchasing extinguishers, especially those to be used on the ladder truck, the manufacturer should be requested to add about six or eight feet to the hose usually furnished. The shoulder straps should be removed, as they oftener prove a nuisance than a help.

STATIONARY CHEMICAL ENGINE.

Upon the same principle as the above described portable fire extinguishers is made a stationary chemical fire extinguishing apparatus, known as the "Champion Stationary Chemical

Engine." It consists of a tank with a capacity of 100 to 1,000 gallons, placed in the basement. From this a stand-pipe ascends from floor to floor as high as needed. On each floor there are hose connections, and hose on a reel. By touching an electric button, or by pulling a wire in connection with the apparatus, the tank in the basement is tipped, and the same chemical action occurring in the portable extinguishers takes place. The stream can be thrown about 75 feet.

The expense incurred in the protection in this manner of every portion of a large institution, with many detached buildings, would be very great, and much more than would be nec-

FIG. 5.—Portable Chemical Engine.

essary for as good protection by other means. The chief objection, however, is that which applies to the use of any form of

indoor *stationary* apparatus: — that is, the liability of the apparatus being inaccessible, when most needed, on account of
smoke or of the location of the fire.   If used, it should not be
depended upon to the exclusion of the more portable appliances.

PORTABLE CHEMICAL ENGINES.

Portable chemical engines are becoming more and more
popular with firemen in cities and towns, as a quick and effectual means of extinguishing fire.   They are particularly desirable
in that they save a great amount of loss to goods easily damaged
by water.   They work upon the same principle as do the portable hand fire extinguishers, and the chemicals used are the
same.   They are made in different sizes, from fifty gallons
capacity up.   A two-wheeled engine of fifty gallons capacity
[Fig 5,] would be a convenient and useful piece of apparatus
for a hospital, especially one not provided with an abundant
water supply, or with inefficient facilities for using it to the best
advantage in case of fire.   A chemical engine should not take
the place of the smaller extinguishers, nor of a *reliable* water
supply, but if the water supply is unreliable, or for any reason
whatever, is not to be depended on, an engine of this kind
should be put into service.   It is light enough to be easily
drawn about by a company of four men, and if properly used
will accomplish much good.

STAND-PIPES.— An iron pipe, connecting with a water-
main, is run up through a building from cellar to attic.   To this
are connected on each floor, several lengths of hose, either
reeled or laid upon racks, with nozzle attached.   Many public
buildings are provided with this system of protection, which is
generally more ornamental than useful, and in case of actual
need usually, if used at all, proves entirely unsatisfactory.
The very cheapest grade of hose, which will scarcely hold
water, is often the kind used, and it remains reeled up or lying

upon its rack for months and years. Even though provided with the best of hose, and pains be taken to keep it in good condition, there are many serious objections to the stand-pipe system, which objections have already been considered.

AUTOMATIC SPRINKLERS.

A system of automatic sprinklers could with advantage be placed in the various laundry rooms, and in work-shops in general. But no matter how perfect the system may be when first put in, if not carefully looked after, it is liable after a time to get out of order. There is not much danger of the sprinklers themselves getting out of order without giving notice of the fact, but through carelessness the supply of water may be rendered inadequate, or it may be entirely cut off. There should be a ready means of determining at any time whether water is being supplied the sprinkler system or not. The main supplying the sprinklers should be in connection with the hydrant mains. Outside the building, in connection with the supply-pipe, should be placed a stop-valve, which should be *sealed open*.

FIRE-HOSE.

A cheap grade of fire-hose is by far the most costly purchase that could possibly be made; that is, if it is intended for use in case of fire. If it is simply to be placed on exhibition for the purpose of deluding the public, then it will answer admirably. Large quantities of absolutely worthless hose are manufactured because there is a demand for an article that is cheap, and which it is not the intention to put to any practical use. Thousands of public buildings, hotels, theatres, manufactories and hospitals are provided with this dangerous stuff. "This hose will not burst under any amount of pressure," said a dealer, and then explained, "because it is so porous and seive-like that it will not hold water under *any* pressure!"

While the use of hose in connection with inside stand-pipes is undesirable, for the reasons already given, if for any reason this must be done, the hose used, in order that it may be of service at all, must be of the very best quality.   For this purpose a *thoroughly reliable* unlined linen hose, that will stand a pressure of 400 pounds, and that will do no more than sweat when water is first forced through it, is to be preferred to any other.   It has been recommended by Mr. John R. Freeman, H. E., that "the manufacturer of unlined linen hose be required to guarantee that the hose bearing his name and trade-mark shall not, upon delivery, burst at a water pressure less than 400 pounds to the square inch."

A thoroughly good quality of medium weight two and one-half inch cotton rubber-lined hose would be the best for use in connection with outside hydrants for fire purposes.   Being wet, it is easily and quickly dried, and an occasional wetting rather improves than injures it ; it will stand the wear and tear occasioned by its use in drills much better than will linen hose, and the loss of pressure in the stream by friction is very much less.

Fig. 6.   Fig. 7.

FIGS. 6 and 7.— Two-ply, seamless, woven cotton rubber-lined hose.

The rubber lining should be very smooth, as the frictional loss due to a rough lining is quite marked.   The durability of the outer surface of hose should also be considered.   It must be so compact as to permit of the least possible absorption of water, or of the introduction into its meshes of dirt or sand.

The *quality* of hose for hospital use should be as good as that used by the fire departments of the larger cities. Before purchasing hose it would be well to consult the chief of some city fire department as to the best grade of hose in use at the time.

Hose, used by institutions situated within or near a city having a fire brigade, should correspond, as to size of couplings, with that used by the city.

HOSE CARTS AND EQUIPMENT.

The hose carts, two to four or five in number, should be *strong*, light, weighing about 600 pounds, and simple in construction, with a carrying capacity of 300 feet of hose. [Fig. 8.]

Fig. 8.—Hose Cart.

EQUIPMENT.—Besides 250 or 300 feet of hose, there should be for each cart a play-pipe and nozzle. The play-pipe should be light, but strong and durable; the thread must correspond perfectly with the thread of the hose coupling. A plain,

smooth nozzle is as good, if not better, than any other,* and is
to be preferred to any complicated affair. One or two spray
nozzles, by the use of which a fine spray of water is thrown

Fig. 9. Combined hose and ladder strap, belt and spanner. As belt.

about the men holding the pipe, would be useful. They could
be carried in the tool-box of the cart, to be used if necessary.

There should be one or two extra pipes and nozzles, to be
kept at the hose-house. Each hoseman should be provided
with a spanner and a hose-strap, or better still, a combination,
as shown in Fig. 9. These spanners and straps should be kept
on the cart when not in use. The tool-box should contain the
hydrant wrench.

LADDER TRUCK AND EQUIPMENT.

The ladder truck should be well and strongly made and at
the same time light enough to be easily handled by eight or
ten men. A poorly made truck would not long withstand
the rough usage it must necessarily receive at the hands of
unskilled men ; whereas, one that is strongly made, of the best
material, would last many years. A good truck for hospital
use would weigh about 1,200 pounds.

*"A ring nozzle of ordinary form, with shoulders 1-16 to ⅛ inch deep
discharges only about three-fourths as much water as a smooth nozzle of
the same size."—John B. Freeman, H. E.

EQUIPMENT.—The ladder truck should be equipped about
as follows: One extension ladder, made up of two ladders, 21
and 18 feet respectively, splicing 36 feet; one extension (16
and 13 feet) splicing 26 feet; one single ladder, 22 feet; one

FIG. 10. Hook and Ladder Truck. Hangers for fire-hats instead of fire-buckets.

roof ladder, 12 feet; two fire axes; two pike poles; four "Babcock" fire extinguishers; four lanterns; ten or twelve hangers
for fire hats; two tool boxes; a crow bar; a shovel, and a
broom; 100 feet of rope for use by hose companies.   If it is
necessary to have a ladder larger than the 36 foot extension,
it should not be carried on the truck, but placed at a convenient point, under cover, near the place where it would be most
needed.

THE HOSE HOUSE.

The hose house should be substantially built and centrally
located.   It should be large enough to contain, without
crowding, the ladder truck, hose carts and chemical engine,

if there be one. There should be a room for the hose-house janitor, and a closet or store-room for supplies of various kinds. It would be a good plan, also, to provide sleeping apartments for those men who compose the chemical company. By so doing, there would be a very material saving of time in getting the truck and chemical extinguishers on the way to the fire at night.

The floor should pitch slightly towards the center, so that water used in washing the apparatus could run into the sewer pipes, made to connect with an opening in the floor. The doors should open outwards, and be of sufficient size to allow of the *easy* passage of the widest vehicle. *The common male ward key should control the door lock*, so that delay need never be necessary in looking for the hose-house key.

If there is no tower for drying hose, racks should be made, long enough to carry a full length (50 ft.) of hose, inclined enough to allow for drainage. If a tower is provided, shelves or racks may be made, upon which extra lengths of dry hose, coiled, may be placed.

Hooks for fire-hats and coats should be conveniently placed. The house should be well-ventilated, and kept warm in winter.

The *hose-tower* is a great convenience, and is necessary for quickly drying wet hose. There should be some means provided for furnishing heat to the tower when wet hose is hanging in it. The distance from the ground to the hangers for hose must be at least fifty-three feet.

CHAPTER IV.

THE FIRE ALARM.

SOME reliable method whereby the members of the Fire Brigade may be quickly and surely notified that their services are required is essential to a proper system of fire protection. Most disastrous results have followed the loss of time between the discovery of a fire and the giving of the alarm. These delays are generally due to the fact that either no system whatever is used, or else the system used is unreliable. Where life and property are in danger from fire, and every second's time is of the utmost importance, there should be the smallest possible amount of time lost in properly notifying those who are to save the life and property endangered.

The notification of the whole brigade should be immediate. Where fire is concerned there is no time for messengers to be sent hither and thither to notify this or that person that a fire is in progress. Neither should the discovery of fire be a matter for investigation by some one in authority before an alarm is sounded. Better that any number of unnecessary alarms be sounded than that the dangers of delay be risked in a single instance.

THE FIRE BELL OR FIRE WHISTLE.

Whatever is used for sounding an alarm of fire, be it bell or whistle, it should never be used for other than fire purposes. There should never be occasion for a moment's doubt in the minds of the firemen when they hear the fire alarm. Each member of the Fire Brigade should know *at once*, when he hears a particular bell or whistle, what is expected of him. The regular alarm-bell or whistle may be supplemented by the blowing of another whistle or by other means, but the ringing of the fire bell or the blowing of the fire whistle should mean but one thing, and should of itself carry definite information to the ears of the firemen.

The bell or whistle should be operated by electricity, being placed either upon the circuit with the alarm boxes and indicators, or upon a separate circuit, the alarm, in the latter case, being transmitted at the office by means of a switch. The safest way would be to place the bell or whistle on the box circuit, so there would be no delay, after a box is pulled, in sounding the alarm. There would be a chance for considerable delay if a separate circuit were used. The mechanism by which the bell or whistle is operated may be so arranged as to prevent an alarm being sounded except when a box is pulled. Any accidental interruption of the current would give notification at the indicator, but would be insufficient to release the mechanism controlling the bell or whistle.

OPEN AND CLOSED ELECTRIC CIRCUITS.

No system of fire alarm should be operated upon any other than a constantly closed electric circuit. With an *open circuit system* there is never any certainty, except at the time of testing and of giving an alarm, whether the system is in good working order or not. The wires may be cut or otherwise injured; the connections at the battery may be loosened or broken, or the battery itself run down, and the system thus rendered useless for the time being, and the probabilities are that the fact of its uselessness would not be discovered until the regular time for a test, or until an unsuccessful attempt had been made to turn in an alarm of fire. The current of electricity passes through the wires only when the circuit is closed, that is, when a test is made or an alarm is sounded.

The current of electricity in the *closed circuit systems* is constant except when the circuit is opened, by accident or design. The alarm mechanism is not operated by the electric current, but is held in check by it, so that *any* interference with the constant passage of electricity through the circuit, be the interference accidental or otherwise, will liberate this mechanism, and an alarm will be the result. The closed circuit system is known to be in order as long as it makes no sound. The open circuit system may or may not be in order when quiet.

THE GAMEWELL RAPID SYSTEM.

of fire alarm telegraph, now in successful operation at the hospital at Kankakee, Ill., would seem to the writer to best meet the requirements of public institutions. This system consists essentially of mechanism held in check by electricity. The clock work signaling mechanism of the alarm boxes is accurately and substantially made, each box with a circuit-breaking wheel, which, in revolving, breaks and closes a closed electric circuit a number of times, corresponding to the number of the box or station. This clock-work is secured in an iron case, or box, with glass or metallic face, through which the actuator of the mechanism projects, to give the means for sounding an alarm. This case of signal mechanism is enclosed within an outer iron case or box, the door of which may be secured by a trap-lock, keys being distributed amongst employes. The Rapid Indicator [Fig. 12] with loud vibrating gong, placed at some central point, indicates the number of the alarm station or box operated. The mechanism of the box is set in motion simply by the pulling down and letting go of the actuator. [Fig. 13.] Before the person operating the box has time to turn about, the pointer of the indicator is thrown to the number on the dial corresponding to the number of the station or box from which the alarm is sent. At the same instant the gongs are vibrating, and if the fire-bell or whistle is included in the circuit, the alarm will be sounded. The system being operated on a closed circuit, any accidental interference with the current would be

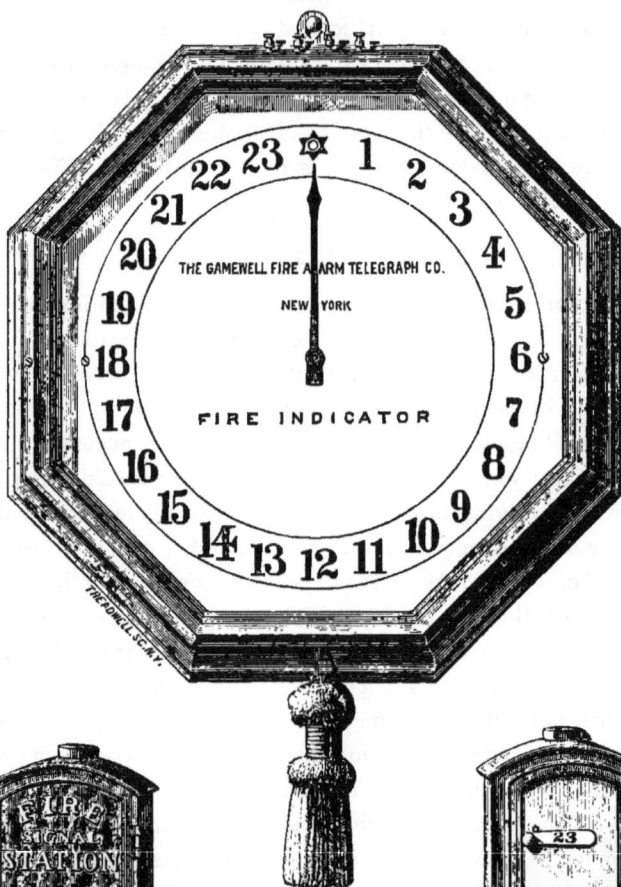

FIG 11.   FIG. 12.   FIG. 13

as quickly known as would the turning in of an alarm of fire.

The system should be put in service by those familiar with the work, so that the necessary care and attention will be given to the adjustment of the various mechanisms, to the wiring, etc. The system should be tested at regular intervals, once or twice each day, and every part of it kept in good order by the electrician.

BOXES OR STATIONS.—The alarm stations [Figs. 11 and 13] should be placed at convenient points, in hall-ways, corridors, etc., if indoors, and on telegraph or electric light poles, if outside, the distance from any given point to a box being not greater than 75 or 100 yards. Where boxes are placed so as to be exposed to inclement weather, or to interference by mischievous patients, it would be well to enclose them in larger wooden boxes. If so arranged, a keyless door could be used on the alarm box, a trap-lock securing the door of the outer box. The numbering of stations should begin at eight or nine, the smaller numbers of the indicator being reserved for registering false alarms. The reason for this is that there would then be less liability of an accidental break in the circuit registering a station number. Accidental interruptions in the current will, in the majority of instances, throw the pointer of the indicator to figure one of the dial, but it will occasionally happen that the pointer will be sent beyond this.

KEYS.—Each officer and employe of the institution should be furnished with a numbered fire-alarm key, which should be kept on the key-ring with the keys in daily use. The Fire Marshal should keep a list of those having keys, the number of the key being placed opposite the name of the holder, so the Marshal may, after releasing a key from the trap-lock of a station, know to whom the key belongs. Holders of keys should be made thoroughly familiar with their use.

THE INDICATOR.—An Indicator [Fig. 12], with vibrating gong attached, should be placed in the main hall, near the executive offices, and one in the hose-house. Near each indicator should be hung a card upon which is printed the number and location of each station.

THE CIRCUIT.—The wire connecting the stations, indicators, etc., constituting the circuit, should be well insulated, and of sufficient strength and conductivity to insure permanence and economy of battery. In buildings it is well, if possible, to have all wires concealed, and wherever within easy reach they should be enclosed in gas-pipe. Where strung on poles with other wires, the fire-alarm wires should be placed above all others, to guard against possible interference with the circuit by being crossed by other wires.

The boxes or stations, indicators, gongs, and fire-alarm bell or whistle, may be included in the same circuit, or the boxes, indicators and gongs may be placed on one circuit, and the alarm-bell or whistle on another, the latter to be operated from the office by a switch placed near the indicator. The single circuit is the safer way. The ordinary gravity battery is to be preferred to any other for operating a closed-circuit system.

THERMOSTAT ALARM SYSTEM.

A *closed-circuit* thermostat system may be advantageously employed as supplemental to the system described above. The thermostats could be placed in attics, basements, and throughout the various work-rooms and shops. The annunciator should be placed in the main office, from whence the alarm by whistle or bell could be given by an electric switch or by turning in the alarm from the nearest alarm station.

CHAPTER V.

THE FIRE BRIGADE.

AFTER an institution is well supplied with the necessary facilities for extinguishing fire, and has also a reliable fire-alarm system, it still cannot be considered as properly protected unless there be in addition some system of organization and drill whereby the apparatus may, when needed, be handled intelligently. There is scarcely any institution, large or small, where it would be impossible or inexpedient to have some form of organization and drill.

Fire apparatus in warehouses, public buildings, hospitals, etc., generally prove useless at the time of a fire because of the very small amount of attention ever paid to the matter by those employed in and about such buildings or institutions. It is not an unusual thing to find hose rotting on the reels, fire-pails empty, or nearly so, and many other evidences of neglect, all pointing to the absence of system. It is rather the exception to find a single person about a public building who is at all practically familiar with the manner of using any of the fire-extinguishing facilities with which it may be provided.

"Premising that my experience is limited to the supervision of the factories which are insured by the Factory Mutual Companies, I beg to say that our experience has been that private fire apparatus is of little or no service unless it is put in the charge of a well-organized private fire department, regularly drilled, consisting of men who are employed in the factories or works; there are few points on which we are more urgent than in promoting such organizations."—Edward Atkinson.

The Fire Brigade should be an essential part of the machinery of fire protection. A great display of fire apparatus is no protection in itself. A few dollars' worth of the simplest apparatus, in the hands of half a dozen brave, cool-headed, well-drilled men, is far better protection than is many thousand dollars' worth of expensive machinery left to run itself, or to be used by excited, untrained men. No matter whether the institution be poorly or generously supplied with the means for extinguishing fire, the necessity for an intelligent use of the apparatus is the same. If fire-pails only are furnished, all the more reason why they should be kept in order, and when used, used promptly and properly. There is no doubt that many large fires in the past could have been prevented had there been at hand, while the fire was in its incipiency, some one experienced in the use of fire apparatus. It is a well-known fact that untrained and inexperienced men lose their heads in the presence of fire, and generally forget about the means close at hand for putting out fire, which they have often seen but never handled.

PLAN OF ORGANIZATION.

The manner of organization of a Fire Brigade in any hospital for the insane will necessarily depend upon the size of the institution, as well as upon the kind and amount of fire-extinguishing apparatus in use. A system is here proposed which would be suitable more particularly for use in the larger institutions. It is intended, however, that what is here offered may be modified to suit the peculiarities of each institution, large or small.

The whole Brigade can be made up from the officers and employes of the institution, and this with but little interference with their other duties. There should always be a certain number in each company, and vacancies filled as soon as they occur. The number of active members must exceed the number actually necessary at any one time, because it is seldom that more than two-thirds of the Brigade can be depended upon to respond to an alarm, except at night, between the retiring and rising hours. During the day and early evening some of the attendants are off duty and away from the hospital, while others are out walking with patients, so that the number who could at once respond to an alarm during the day would be much smaller than could be depended upon at night.

Attendants who are members of the Brigade should be so placed in regard to position on wards as to allow of as many as possible leaving on an alarm of fire; that is, the bunching of firemen on wards should, if possible, be avoided. When this cannot be avoided there should be a system of relief, whereby attendants, not members of the Brigade, could relieve those who are.

The HOSPITAL FIRE BRIGADE should consist of :—

A Chief Marshal.
Two Assistant Marshals.
An Engineer.
Assistant Engineer.
An Electrician.
Two or more Hose Companies, of six men each.
One Hook and Ladder Company, of ten or twelve men.
A Chemical Company, of five or six men.
A Life-Saving Corps.
A Hose-House Janitor.

CHIEF MARSHAL.

The Chief of the Brigade should be chosen from the officers of the institution, preferably from the medical staff. This for several reasons: better discipline can be maintained; a better opportunity is afforded a physician for becoming thoroughly acquainted with all parts of the institution ; he has a better opportunity for instructing employes from time to time concerning matters relating to the prevention of fire, etc.

The Chief should be active, strong, in good health, brave, cool-headed, and possessed of tact and judgment. If a practical fireman be not employed to assist in the organization of the Brigade, the Chief would do well to familiarize himself in fire matters by a visit to a well-organized city fire department, to one already in successful operation in some other institution, or to both.

As the responsibilities of the Chief are many, much authority should be vested in him, and it ought to be understood that at the time of a fire or drill, in matters pertaining to the fire or drill, the Chief of the Brigade, acting for the Superintendent, is highest in authority. If this is not done, great confusion is apt to occur, which might prove quite serious. The Chief should appoint the officers and members of the Brigade ; personally conduct the drills, and give instruction ; frequently examine all apparatus, and see that it is at all times in good working order; inspect, from time to time, every nook and corner of the institution, so that he may intelligently direct the movements of the firemen at a fire, day or night. All measures for the prevention of fire must be under his supervision, as well as the means for extinguishing it. It should be his duty, under the direction and with the approval of the Superintendent, to frame certain rules and regulations regarding the prevention of fire, and he should see to it that these are rigidly adhered to. These rules should be printed, and placed so that employes in general could often refer to them. This, however, is not enough. Each employe should be thoroughly instructed as to the meaning of every rule, and the Chief should, by frequent inspection, see that the rules do not become a dead letter. As much importance should attach to the prevention of fire as to any other part of the Chief's duties.

A record should be kept by the Chief of all matters of interest in relation to the doings of the Brigade. Minutes should be made of all drills — giving the names of the companies taking part, the time made, etc. Also an account of each fire, no matter how small, time of alarm, cause of fire, how discovered, how extinguished, damage done, etc.; irregularities existing, whether remedied, and how. A record of this kind would prove not only interesting but instructive as well.

Upon an alarm of fire, day or night, the Chief should proceed directly to the location of the fire, so that he may know exactly what is best to be done upon the arrival of the apparatus, as well as to be able to direct, at the earliest possible time, the movements of those in the vicinity of the fire. Upon the arrival of the Brigade, the Chief will generally have formed his plans, and be able, without delay, to give the proper orders to his Assistants.

After a fire, false alarm or practice, the Chief should return to the hose-house with the firemen, and before dismissing them call their attention to mistakes made, and congratulate them upon good work done.

Certain signals should be decided upon between the Chief and his Assistants, either with whistles, or lanterns, or both.

ASSISTANT MARSHALS.

Even in the smaller public institutions it would be best to have two Assistant Marshals, for the reason that in the absence of one of the three there would be two left to conduct affairs in case of necessity. The Assistants should be chosen from the medical staff or Supervisor's or business departments, if possible. They should assist the Chief in the details of the management of the Brigade, and what has already been said concerning the duties and responsibilities of the Chief in a large measure applies to his Assistants, for in the absence of the Chief one of the Assistants must be at the head.

Certain *special* duties should be assigned to each Assistant, and these duties ought to be thoroughly understood by all.

The *First Assistant Marshal* should go direct to the fire, upon an alarm, and take charge until the arrival of the Chief, and after that assist him in his plans and take orders from him. It should be his special duty, upon the arrival of the Brigade, to see that the different companies are properly placed without delay, and to direct their movements in accordance with the general plans of the Chief. Much confusion would result if no one were to meet the different companies upon their arrival.

The *Second Assistant Marshal*, upon an alarm of fire, should go to the hose-house and take charge there. He should assist the Captains in forming their companies, give them the location of the fire, and direct them as to the best road to take. After doing this he should go to the fire, assist in the placing of companies, or report for instructions to the Chief.

ENGINEERS.

The Engineer of the institution should be, as well, the Engineer of the Fire Brigade. His duties would include the care of fire-hydrants and mains, the management of the engines and pumps in case of fire or drill, etc. On an alarm his place should be at the pumps.

The *Engineer's Assistant* should be of the engineer's department, and his duties would be to assist the Engineer, or, in his absence, take his place.

Special rules for the guidance of those connected with the engine and pump-rooms should be framed and conspicuously displayed. Everything about the Engineer's department that has to do with the furnishing of water in case of fire should be in such shape that the least possible amount of delay would be had in giving the necessary amount of fire-pressure at a moment's notice.

The Engineer should frequently and regularly test fire-hydrants, winter and summer, and keep them in good repair. He should report to the Chief at once the fact of any hydrant or main being out of service, even though it be but for a few hours.

ELECTRICIAN.

The care of the electric fire-alarm should, if possible, be entrusted to one experienced in the handling of electrical appliances. In those institutions where the electric light, watch-clocks and telephones are in use, it is an easy matter to have a practical electrician for the fire-alarm apparatus. In the absence of such a person, however, some one of the staff could, by a little study and trouble, sufficiently familiarize himself with the fire-alarm system to be able to manage it and keep it in order.

The Electrician should make a thorough test of the fire-alarm once or twice a day, at stated hours. This could best be done by turning in an alarm from some fire-alarm station—a different station or box each time. The test could be made so as to prevent a general alarm being sounded. After each test he should see that everything is in readiness for the proper working of the system. Any irregularities found to exist must be attended to at once, and if, for any reason, the system, or any part of it, is thrown out of service, even for a short time, he should report it at once to the Chief, so that other means for giving an alarm could be substituted without delay.

Immediately after an alarm of fire the Electrician should see that the system is put in shape for another alarm ; that is, the indicators, drops, alarm-bell, or whistle, etc., should be re-set.

HOSE COMPANIES.

The number of hose companies to be formed will depend upon a variety of circumstances : the size of the institution, the amount of hose and number of carts in use, and the number of

men available.   There should, however, never be less than two
companies, even for the smaller public hospitals, while a
greater number than four or five would not be necessary for the
largest.

Each company should consist of six men, as follows:
a *Captain*, *Lieutenant* or *Pipeman*, a *Hydrant-man*, and three
Linemen.

DUTIES OF CAPTAIN.—Upon the Captain of a Hose Company depends, to a great extent, the success of the company's
work. He should fully understand the importance and respon-
sibility of his position, and seek not only to work well himself,
but also to have his company know and respect him so well
that every order he may give will be carried out quickly and will-
ingly.   Upon arriving at the hose-house, in case of an alarm,
the Captain's first duty should be to help the Assistant Marshal, if one be present, in forming companies and starting them
on the road.   The Captain, in taking possession of a cart,
should get his company together by calling, " This way, No.
—."   Upon leaving the hose-house, the Captain should lead,
either running ahead or taking his place at the head of the
cart, and give orders which way to go and to what place.   His
orders should always be given so that they will be distinctly
understood by the members of the company. He should never
be afraid of speaking loud in giving orders, for in the excitement and noise, an order given in an ordinary tone of voice will
be very likely unheard or misunderstood. Very few orders
will be necessary in directing the movements of a well-drilled
company.   On the way to the fire the Captain should form his
plans as far as possible, so that he may know how to direct his
company. Upon arriving at the place of fire, if no orders are
given from a Marshal, the Captain must use his judgment in
deciding to which hydrant to connect, and what to do after the
connection is made. If after laying a number of lengths of
hose no orders are given, the Captain should leave his company,

seek a Marshal, or find out where the fire is, and act according to his best judgment in directing his company what to do. In the absence of the Marshals, the first Captain arriving at the fire should take temporary charge. The Captain must direct all the details of the work done by his company. He should not confine his attention to the pipe end of the line, but be sure that all is well from one end to the other. Before leaving the hose-house, after returning from a fire or practice, the Captain should carefully examine his cart, to make sure that everything is there and in its place, that is, that his cart is "in service."

DUTIES OF LIEUTENANT OR PIPEMAN.—In the absence of the Captain of a hose company the Lieutenant or Pipeman will take charge until the arrival of the Captain, or the appointment of one by a Marshal. He will give the pipe to another member of the company, if he thinks best. Immediately upon arriving at the hose-house the Pipeman should secure the hat and coat set apart for his use, put them on as quickly as possible, and then take from the cart the pipe and a spanner. On the way to the fire he should run ahead of the cart to assist the Captain in leading the way, or in the absence of the Captain, to take the lead himself, except in case the company is small, or the roads bad, when he should take his position on one side of the pole of the cart. When sufficient hose is unreeled, he should, with the assistance of one other, connect the pipe to the hose, and then go to the fire where ordered.

DUTIES OF HYDRANT-MAN.— The Hydrant-man (and no matter how small the hose company, there must be one), upon leaving the hose-house, with his company, must first secure a hydrant wrench from the box on the cart, and then take his position *behind* the cart, ready to secure the end of the hose when near the hydrant to which he is to connect. After making the hydrant connection, he should place the wrench on the hydrant rod, and stand ready to turn on water when he receives the order to "play away." He must remain at the hydrant until otherwise ordered. His position is a very important one, and he should be always on the lookout for orders. He must be especially careful not to act upon the suggestions or orders of outsiders. Returning from a fire or drill, he should see that the hydrant is left in proper shape, and be sure he takes his wrench with him.

DUTIES OF LINEMEN.—After reaching the hose-house, and taking position at one of the carts, Linemen must take the direction given by an officer, usually the Captain of their company, and assist in getting the cart to the fire as quickly as possible. They should always pay close attention to the orders given by the officers, and execute them thoroughly. If the regular Hydrant-man is absent, one of the Linemen must take his place. It is the especial duty of the Linemen to unreel hose, carry it to any point ordered, and see that it is properly laid. After this their duties are generally with or near the pipeman.

HOOK AND LADDER COMPANY.

This company should be composed of ten or twelve stout and agile fellows. At a fire many things are to be done by the Ladder Company, and often several things at the same time. Each order should be executed quickly and smoothly. There must be no misunderstanding of orders, and there should be such a thorough understanding between officers and men as to avoid all confusion in the carrying out of any order that may be given. The exact work to be done by the Ladder Company will depend in each case upon the location, progress and extent of the fire. The raising of ladders, opening of doors and windows, tearing away plastering, removing partitions, flooring, and assisting in the saving of life, are a few of the duties a ladder man may be called upon to perform. Arriving at the location of the fire, the Captain of the Hook and Ladder Company will receive instructions from one of the Marshals, and he will give such orders to his company as will be necessary for the execution of the plans of the Chief or Assistant. While taking an active part in the work themselves, the officers of the Ladder Company should direct the details of what is done.

CHEMICAL COMPANY.

Where the portable (Babcock) hand fire extinguishers are used, the better arrangement would be to have four extinguishers, of six gallons capacity each, placed on the ladder truck, two on either side. [Fig. 10.]

The Chemical Company should consist of a Captain, Lieutenant, and from three to five men, thoroughly drilled, and each made practically familiar with his work. The Captain must take his orders from a Marshal, if one be present, and direct in general the work of the company. In the absence of a Marshal, he must do as his judgment suggests.

If a *Chemical Engine* be in service, the organizantion of the Chemical Company would, of course, be different from that above suggested. The smaller extinguishers could still be carried on the ladder truck, and members of the Ladder Company be drilled in their use. The number of men necessary to compose a Chemical Engine Company would depend upon the size of the engine. If a fifty-gallon engine, five men would be sufficient, but for a 100-gallon engine a larger number would be required.

As the early arrival at the fire of the Chemical Company is highly essential, the members of the company should, if possible, have their sleeping quarters at or near the hose-house.

AT THE FARM.

In those institutions where the farm buildings are quite a distance from the main ˙structures, there should be formed, from among the farm employes, one or two companies, and they should be supplied with a hose-cart, a few ladders, axes, poles, lanterns, etc., and two or more chemical extinguishers. The general plan of the organization of such companies would be essentially the same as described above. Fires are as liable to occur in farm buildings as elsewhere, and the having at hand of the means for extinguishing small fires, or of controlling or holding in check more serious blazes, pending the arrival of help, is a very desirable thing.

LIFE-SAVING CORPS.

The Life-Saving Corps should consist of a Captain, Lieutenant, and from five to ten members, the membership to be made up from the administrative offices principally. The Superintendent of the institution, or one of the Assistant Physicians, ought to be Captain, the other members being physicians, the druggist, clerks, and male and. female supervisors.

At frequent intervals the Life-Saving Corps should inspect various parts of the institution, especially those where, in the event of fire, their services would be most needed. They should, at these visits, study well the means of egress from the wards and dormitories; make themselves familiar with the fire-escapes by using them; and determine, as nearly as possible, as to the manner in which certain wards and dormitories could be emptied of patients quickly and safely. They should know where are located the feeble, the sick and the otherwise helpless patients, and as well those who would probably resist rescue, or in other ways cause trouble. Much may be done, at these inspections, in the way of forming general plans of action in anticipation of the occurrence of fire in different parts of the institution. The knowledge gained by the Corps at these visits will enable them to accomplish a great deal in a very short time, if they are ever called upon to attend a fire.

HOSE-HOUSE JANITOR.

It should be the duty of the Hose-House Janitor to care for the apparatus; to keep it clean and in order; to see that the wheels of the truck and carts are regularly greased; to attend to the drying of hose after practice, etc. He must have strict orders regarding the lending of hose, wrenches, lanterns and other things belonging to the paraphernalia of the Brigade. He should also keep an inventory of everything connected with or belonging to the Fire Department.

The lanterns on the truck and carts should be lighted at dusk and kept burning all night. In case of an alarm of fire the Hose-House Janitor should open the doors and assist in getting the apparatus out of the house. If a member of a company, he must join it as it is leaving.

FIRE HATS, RUBBER COATS, BOOTS, ETC.

The members of the different companies should be provided with fire-hats, rubber coats, belts and boots. The fire hats, coats and belts should be hung up in convenient places at the hose-house, either on racks or upon the apparatus, where they can be easily secured by the firemen. The boots, being used only at night, should be kept at the fireman's bedside, with his pantaloons, coat, hat, etc.

"BUNKERS."

Pantaloons, drawn on over a pair of boots — either rubber or leather,— the former slipped down over the latter, left open in front, form what is known by professional firemen as "bunkers," a very simple but useful, convenient and time-saving device. [Fig. 14.] These should be set alongside of the bed, and upon the floor near them may be placed an old coat, in the pockets of which are kept a cap or hood, mittens, keys, a silk handkerchief, as a respirator, and a box of matches. A few trials at getting out of bed, into the "bunkers," and out of the house in the dark, will prove to any one what an immense saving of time can be secured by their use. In warm weather the coat could be put on after leaving the house. Much time and clothing can be saved if each fireman would adopt such a rig and keep it handy.

Fig. 14. "Bunkers."

CHAPTER VI.

DRILLING OF THE BRIGADE.

"A drill for an organized body is as essential as the body itself, and the more diversity it contains, as the members gradually become skillful, the less the monotony, hence an inducement to greater proficiency. It has been fully demonstrated that a few skilled and disciplined men are of more service than a host of men untaught. To be able to perform an act, the fundamental principle is to know how. And to acquire that knowledge, assiduous study and constant practice in all the branches pertaining to the service is an absolute necessity."— D. J. Swenie, Chief of Chicago Fire Department.*

The importance of the drilling of the Fire Brigade cannot be overestimated. The less actual fire duty an organization is called upon to do, the more drilling and practice is necessary in order that each member may be ready when the time for action comes. This is especially true where the *personnel* of the Brigade is constantly changing. It is impossible for firemen to learn how to use fire-extinguishing apparatus and to work in harmony with each other, simply by reading, or by being told, or by watching others. Each fireman must have *practical* knowledge of his work, and this knowledge can only be secured by dint of *frequent and thorough practice with the apparatus.*

If this matter of frequent drill be neglected, the Fire Brigade will without doubt become more and more inefficient, and it will be found, when too late, that the apparatus was not in order, or that the firemen did not know what to do.

* Report National Association Fire Engineers, 1887.

Just as soon as the Brigade is organized, a system of drill should be instituted that will give to each fireman as thorough and as practical a knowledge of his duties as possible, the drills being kept up throughout the year. Much attention must be given to the drilling of separate companies, and to each member thereof — at first, when the organization is new, and from time to time thereafter as the membership of the different companies changes. At first single companies only should be drilled, but after a little, two or more companies may be drilled at the same practice, either alternately or together. Each fireman should become familiar with the work of every other member of his company, so that he could fill another's place upon a moment's notice.

CONTESTS.—An occasional friendly contest between the different Hose Companies should be encouraged. Instead of contests, the Ladder Company could have an occasional trial of speed, competing against former records made. Suggestions as to these contests will be found under the appropriate headings farther along in this chapter.

SMUDGE FIRES.—It is of the greatest importance that firemen should become more or less accustomed to working in a smoky atmosphere, for no matter how well drilled they are in handling apparatus, they would find it very difficult to do their work properly at a fire were it necessary for them to inhale, for the first time, hot stifling smoke. It is suggested, then, that smudge fire drills, as hereinafter described, be occasionally had.

FALSE ALARMS.—Once in six weeks, or oftener, a *false alarm* should be sounded. Besides being an excellent way in which to test the efficiency of the Brigade in the matter of responding to an alarm, it is the very best method in which to prevent the occurrence of a panic among patients and others when there really is a fire. They soon become accustomed to hearing the alarm and seeing the firemen at work, and pay no more than ordinary attention to what is going on. It can very readily be seen of what inestimable advantage this fact would be in case of fire. The firemen also are benefited by these false alarms. Ignorant that the alarm is false, they respond as in the case of fire. It accustoms them to being called without previous warning, night or day, and gives them, in a general way, actual fire practice. It gives the officers of the Brigade an opportunity for watching for and correcting mistakes made by individuals or companies, and for knowing what could be expected of the Brigade at a fire. The false-alarm drill can be so conducted, generally, that the firemen need not know, until the drill is over, that they were not in attendance upon a genuine fire.

OBEDIENCE TO ORDERS.—What is said in regard to the necessity of firemen obeying the orders given them by the proper officers at a fire applies with equal force to their so doing at drills. Poor discipline at a drill means poor discipline at a fire. If the habit of obeying orders be well formed at practice, then there need be no fear of plans miscarrying at a fire on account of disobedience or tardy compliance.

VOCABULARY.—Members of the Brigade should familiarize themselves with the meaning of the expressions employed by the officers in giving directions. Following is a partial list of the more common phrases used, with a few words in explanation of each:

"*This Way*, ———."—This order is always followed by the name of the company or individual addressed, and means that said company or individual is to approach or follow the officer giving the command.

"*Lead in*," "*Up*," "*Down*," "*Forward*," "*to Right*," or "*to Left*," indicates the place and direction in which it is desired a company is to go, with apparatus, the order being followed by the name of the company, as well as by the place or distance it is intended the company is to go.

"*Light Up on that Hose*" is generally an order from the Pipeman to the Linemen to get the hose forward.

"*Play Away*" is ordered by an officer to a Hose Company when water is wanted and everything is in readiness for it. It is the signal for the Hydrant-man to turn on water. It is also the proper order when a stream from an extinguisher is wanted.

"*Shut Down.*"—This means, for a Hose Company, that water is to be turned off at the hydrant; for the Chemical Company, that the stream is to be turned off at the nozzle.

"*Back Out.*"—This order is given to a company when it is desired that the company is to back out of or retreat from a room or building. The retreat should be made in good form, but *at once*. Unless otherwise ordered, the apparatus, hose, extinguishers, etc., should be taken along.

"*Back Down.*"—This order is similar to the above, except that it applies to backing down a ladder or retreating down stairs.

"*Take Up.*"—This is usually given after the fire is extinguished, and means, literally, to prepare to return to quarters.

'*Disconnect.*"—This order is given to the Hydrant-man just before a Hose Company commences to reel hose in "taking up."

"*Go Home.*"—The companies designated, upon receiving this order, return to the hose-house.

THE DRILLS.—A careful study of and frequent reference to the drills now about to be described, will, it is hoped, enable the inexperienced members of the newly organized Brigade to become skillful in the handling of the apparatus. Each step is described somewhat in detail, for the reason that nearly all of the members of hospital fire organizations will have had but little if any experience, and that with some such guide the work will be made much easier than were they to attempt to work out the details without it. It is intended that these drills, taken collectively, will cover, as nearly as possible, the more important steps in the fighting of fire. These being mastered, modifications of each will be suggested in the minds of the officers.

Where previous notice of a drill is given, firemen should put on their "bunkers," for the purpose of saving their clothing from wear and tear and dirt.

Fig. 15. Cart "in Service."

HOSE COMPANY DRILLS.

GENERAL CONSIDERATIONS.— The following Hose Company drills are planned upon the supposition that the carts have reeled upon them 250 feet of hose, and are well supplied with wrenches, spanners and hose-straps. Each one of the drills may be modified as to the number of feet of hose laid, distance run, etc., but as a general thing 250 feet of hose should be used. In the drills where water is used fire-pressure should be put on.

The washing of windows and sprinkling of lawns is very good practice during the summer months, in addition to the regular drills. Firemen thus become accustomed to handling hose and pipe while pressure is on.

LAYING OF HOSE.— Hose should be unreeled in as direct a line as possible from the hydrant to a point opposite the door or window to be entered, or ladder to be climbed. The cart must not be taken beyond this point, but as much more hose as is needed unreeled. As the couplings drop to the ground they are liable to become indented by striking upon stones or other hard substances; this must be borne in mind, and in laying hose over stony ground or hard pavements, linemen should try and have the coupling fall easily. A slight dent in a coupling is sufficient to impair seriously its usefulness.

Sharp turns should be avoided, as when water is forced through the hose a kink will form at the turn and either burst the hose or weaken it very much. Short turns also interfere with the free flow of the water, by increasing the friction. In going through doorways, halls, and up stairways, where turns are frequently necessary, the widest possible curve must be given the hose. In passing hose over window-sills a sharp bend is liable to occur unless care be taken to prevent it. The same careful attention to this matter is necessary in manipulating hose, changing its position, etc., after water has been turned on. In carrying hose up ladders the hose-strap must be used to secure the hose to the ladder, especially near the top, if the hose is laid so far; if not, as near the pipe as possible. The strap, looped about the hose and hooked upon a ladder-rung, will support considerable weight. The hose must lie upon the rungs midway between the sides of the ladder. The most usual point for hose to give way is just behind the coupling to which the pipe is connected. This is because of the sharp bend given to the hose at this place in practice, both wet and dry. In laying the pipe down for any purpose, as is often done during drills, it should be placed so that a kink or sharp bend will not form behind the coupling. This sharp bending of the hose in one place many times will very soon crack the rubber lining, and it is not long then before the hose must be cut down.   Care in this respect will save hose for years that carelessness might ruin in a few weeks.

Sometimes it may be desirable or necessary to have a lead of hose in a fourth or fifth story, or upon the roof, where it is not convenient or possible to have it taken by stairs or ladders.   In such a case the rope carried for this purpose on the ladder truck should be used.   The pipeman and another should take the coil of rope to the place where the lead is desired, retain one end of the rope, and throw the other to the firemen below, who fasten it to the hose, and it is drawn up.

MAKING AND BREAKING CONNECTIONS.— Great care is necessary, in coupling together two lengths of hose, in coupling hose to hydrant, or pipe to hose, to avoid crossing the thread. The male coupling should be held firmly while the female coupling is being fitted and turned into place.   A very slight turn of the female coupling to the left, until it has caught the proper thread, is a very good thing.   Then by making two or three full turns to the right the connection is completed.

In order to make a firmer connection than this, it may be necessary to use spanners, one being fitted to each coupling, in such manner as to exert force in opposite directions. While the spanner on the male coupling is held firmly, the other is turned to the right.   In breaking connection this manner of procedure is reversed.

THE PIPE AND NOZZLE.— Only the ordinary play-pipe and nozzle should be used, unless an exception be made in favor of the spray nozzle, which is of use in fighting a very smoky fire at close range.   Shut-off nozzles would prove a source of danger in other than experienced hands.   The one important rule for the pipeman to bear in mind, in the use of the pipe, whether water is being used or not, is *to keep the pipe in as straight a line with the hose as possible.*

TAKING UP.— The pipe having been removed, and the

connection at the hydrant broken, the hose is ready for reeling upon the cart. It is important that the connection at the hydrant be broken first, in order that the water remaining in the hose may find a ready escape as it is forced along during the process of reeling. With the ordinary village cart it will require four men to reel the hose properly.

The cart is backed up to the pipe end of the line of hose. The fireman whose duty it is to guide the hose as it is being reeled, standing behind the cart, the hose to his right, secures the end of the hose to the reel within a foot of the wheel on the right. This can be done by tying about the reel a stout cord or leather strap, the ends being left long enough to allow of being passed around the hose behind the coupling, and tied into a bow-knot. Securing the hose to the reel in this way is to prevent the too sudden coming off of the coupling in unreeling rapidly, this being especially likely to occur where there is no brake-attachment. Having fastened the hose in this manner, the reel is turned forward by two firemen, one on each side of the pole, in front. The fireman guiding the cart pushes slowly backward along the line of hose. The fireman who guides the hose upon the reel should strive to have it lie even and close. He must hold it firmly as it is being reeled, winding it from side to side, never allowing it to twist nor to hang loose. While greater care is *necessary* in reeling dry hose than when it has been wet, as the latter, on most occasions, will be taken from the cart at the hose-house, still, the best plan would be to reel hose always in the most careful manner. The tighter and smoother hose is reeled the better will it unreel. After the hose is all reeled, the reel must be made stationary, the end of the hose hanging over the back of the cart. A hose-strap fastened to the cross-bar of the cart in front, and another to the bar behind, the other end of each strap being caught upon spokes of the reel, will hold it firmly in place.

THE CARE OF HOSE.—If allowed to remain upon the cart,

wet hose (cotton) is liable to become quickly damaged. For this reason wet hose should be removed from the cart as soon as it is returned to the hose-house, and dry hose substituted. In making this change, the requisite number of lengths of dry hose should be *first* gotten ready, the couplings and washers examined, and the threads treated with a very little tallow or mineral oil. The wet hose is then unreeled, the couplings being broken, or disconnected, as each length comes off. The wet hose is placed at one side, and the dry hose reeled on. The male coupling or nozzle end of a length of hose is fasten-ened to the reel, and the hose is wound snugly into place. Another length is brought, connected with the first by at least three turns; this in turn is reeled up, and the remainder is treated in the same manner. The cart being made ready for service, the wet hose is hung up in the tower or laid upon racks, to dry. In cold weather the drying should be hurried by admitting hot air to the tower. If the hose is dirty it should be brushed off with a broom after drying. Then it should be lowered from the tower, and laid, either in coils or at full length, upon racks provided for this purpose, where it can be gotten at more quickly, if needed, than if left in the tower. It is well to change hose that has remained on the reel more than three or four weeks, substituting for it that which has been exposed to the air. In winter the hose should be kept warm. The foregoing applies to the care of cotton rubber-lined hose.

HOSE DRILL NO. I.

Run 100 yards to hydrant, connect, lay 250 feet of hose, attach pipe. "*Take up.*"

This drill should be repeated several times, very slowly at first, so that each step may be thoroughly learned. After the general order is given by the Chief to the Captain, the latter is to give all the orders as to the details of the work. At the starting point, 100 yards from the hydrant, the firemen should be in the following positions :   The Captain a little in advance; the pipeman alongside the cart, pipe in hand, two linemen at the pole; one lineman and the hydrant-man behind, to push, the latter, hydrant-wrench in hand, standing nearest to the side next the hydrant.   When within fifty feet of the hydrant the hydrant-man begins to unreel hose, keeping hold near the end, at the same time moving towards the hydrant. As the cart passes the hydrant, he will take one turn and a half about it with the hose, placing his foot upon it, that it may not be pulled away from him by the others.   The cap of the hydrant is now removed, and by the time this is done he may safely remove the hose from about the hydrant and make the connection, attaching the hose coupling to the hydrant by at least three turns.   The wrench, which is a combined spanner and wrench, is then fixed upon the hydrant-rod, the hose is given a good free curve, to avoid kinking, and the hydrant-man is then ready to turn on water if the order to "play away" be given. The others, running with the cart, do not stop at the hydrant, unless the hydrant-man fails to secure the hose about the hydrant.   After passing this point the pipeman takes his place behind the cart.   The lineman behind must see that as the hose is unreeled, the couplings do not strike heavily upon stones or concrete walks.   As the last length is playing out, the runners must slow up, stopping the cart before the last fifteen feet are unreeled.   The lineman at the rear takes the end of the hose as it comes from the cart, after it is released from the reel, and the others take the cart to one side.   The rear lineman then straddles the hose, holding the coupling firmly in both hands, while the pipeman attaches the pipe, taking at least three turns, either with his hands alone or by the aid of a spanner.

This completes the first part of the drill.   The hydrant-man should be in position, ready to turn on water, by the time the pipe is in place.   No water is to be used in this drill, however. After a moment's rest, during which the Chief should call attention to mistakes made, the order to "take up" should be given. This must be done as previously described.

HOSE DRILL NO. II.

Same as first part of No. I; then, break and make all connections, including pipe and hydrant.

After laying 250 feet of hose, as in Drill No. I., the pipe is placed upon the ground. Two men are selected to break and make all connections, which they will proceed to do as follows: while one straddles the hose facing the pipe end, holding the male coupling firmly in both hands, the other, facing the hydrant end, breaks the connection by turning the pipe to the left, using a spanner, if necessary, for the first turn or two. Dropping the pipe and hose to the ground, the men proceed to the next coupling, where the same positions are taken and the connection broken in the same manner. Each connection is broken in this way until the hydrant is reached. One fireman will disconnect, with spanner, the hose from the hydrant. The other fireman now makes the connection, with spanner, and they proceed to make connection at couplings, taking the same position as in breaking. Three full turns should be taken. The same two should then go over the drill once more, reversing positions. In this manner all the members of the company should make the drill.

HOSE DRILL NO. III.

Run 100 yards to hydrant, connect, lay 250 feet of hose, attach pipe, taking hose into building. " Back out." " Take up."

Instead of laying the hose in a straight line for 250 feet, *the cart must be taken only as far as a point opposite where it is intended the building is to be entered.* The balance of the hose is unreeled at this point, the pipe attached, and the hose or lead is carried into the building. The pipeman leads the way, holding the pipe in such manner as to prevent a sharp bend in the hose behind the coupling. The hose may be thrown about his shoulders in such a way as to prevent this kinking, as well as to give him a better purchase upon the hose in going forward. A lineman should follow the pipeman not further than ten or twelve feet, the hose being carried by the side or thrown over a shoulder. The other two linemen should follow, at a greater or less distance apart, according to the amount of hose to be carried in. The linemen should not only assist in carrying the hose, but endeavor to keep it straightened out, to make it lighter for those in the lead, prevent sharp turns about door-ways, stair-casings and the like, and to have the hose laid in such shape that it will be ready for water by the time the pipeman is in position. After this the lineman next the pipeman goes forward to assist with the pipe, the others remaining along the line to be ready to pass the order to the hydrant-man to "play away," if such order be given. This order will not be given in this drill, however. The order to "take up" being now given by the chief, the Captain shall first order the company to "back out." The hose is taken out of the building in about the same manner in which it was carried in, and the pipe removed. The order to "disconnect at hydrant" is given at the same time. The hose is then reeled in the manner already described.

In this drill a new point should be selected for each practice, and different buildings, and various parts of each building entered — basements, the various floors, attics,— in fact, wherever the 250 feet of hose will reach, without the use of ladders.

HOSE DRILL NO. IV.

Run 100 yards to hydrant, connect, lay 250 feet of hose, attach pipe, take lead up ladder to second or third-story window or to roof. "Back down." "Take up."

This drill should be practiced at the time the Ladder

Company is having its Drill No. II. The steps of the drill are the same as in No. III. until after the pipe is attached, the cart, of course, being taken only as far as a point opposite the ladder, the remainder of the hose being unreeled at that place. The pipeman, throwing the hose over his shoulder, proceeds to climb the ladder. A lineman must follow close upon the pipeman, to relieve him of as much weight of hose as possible. The other linemen assist in getting the hose up the ladder, their positions depending upon the length of ladder, and whether the hose is to be taken part way up, or to the roof. The same careful attention in regard to avoiding sharp turns, kinks, etc., is as necessary here as in No. III. In any case, the hose must be secured to the ladder by a hose-strap, not far behind the pipeman, if he is to remain upon the ladder. If he takes the lead of hose upon the roof, the strap is to be used near the top of the ladder. When there is an excess of hose at the foot of the ladder, it must be laid about in easy curves, so as to cause the least amount of resistance to the flow of water. A lineman should always remain very close to the pipeman to lend assistance in holding the pipe when water is used, no matter where he may be stationed. The remainder of this drill needs no explanation, provided the Company is familiar with No. III.

In climbing a ladder, hosemen should remember to grasp the sides of the ladder, the feet being placed upon the outer extremities of the rung.

HOSE DRILL NO. V.

Two companies — two carts. First Company — Run 75 yards to hydrant, connect, lay 250 feet of hose, couple to second Company's hose. Second Company — Run 75 yards from hydrant, lay 250 feet of hose, attach pipe. "Take up."

The two companies start at the same time, one (which for convenience sake we will designate No. 1) from a point 75 yards on one side of the hydrant, the other (No. 2) from a point 75 yards the other side. No. 1 runs 75 yards, connects to hydrant, and lays 250 feet of hose in the same manner as for Drill No. I. Instead of attaching the pipe, however, the pipeman or a lineman assists the hydrant-man of No 2 in connecting the two leads of hose. After running 75 yards, No. 2 commences to unreel, the hydrant-man remaining to assist in connecting with No. 1. The 250 feet of hose being laid, No. 2 finishes by attaching the pipe. The three attachments, hydrant, middle coupling, and pipe, should be finished at about the same time. In "taking up," the order of procedure should be the same as for single company drills, each company working for itself.

HOSE DRILL NO. VI.

Same as Drill No. I., with water at fire pressure.

This drill, and those following, are useful in accustoming the firemen to handle hose and pipe while water is flowing at fire pressure. The order to "play away" is given as soon as the pipeman is in position. In "taking up," the order to "shut down" is first given, followed by the order, "disconnect at hydrant."

HOSE DRILL NO. VII.

Same as Drill No. IV., with water at fire pressure.

In this drill the stream is to be directed *away* from the building — the reverse of what would be necessary in case of actual fire. In directing the stream from a position upon the ladder, the pipeman must be sure the hose is properly supported by hose-straps, and that a lineman is at hand to assist him. No attempt should be made to go up or down ladders with a stream on, except for a few rounds, or when it is absolutely necessary.

HOSE DRILL NO. VIII.

Same as Drill No. III, with "smudge" fire.

After laying the hose, as in No. III., the lead is carried

into a fire-proof basement, where a "smudge" fire has previously been prepared. This can better be done while the Ladder and Chemical Companies are having a similar drill. No water is to be used, the Chemical Company being sufficient to extinguish the fire. The "smudge" will be described under Chemical Drill No. IV.

LADDER COMPANY DRILLS.

GENERAL CONSIDERATIONS.—The suggestions and drills here given are intended for a company consisting of ten or twelve men, the truck equipped with ladders about as follows: One 36-ft. extension; one 26-ft. extension; one 22-ft. single; one 12-ft. roof ladder. In addition the truck is supposed to be provided with two axes, two pike-poles, etc.

It is of the greatest importance that each truckman should be able to take any position in raising or lowering any of the ladders. For this reason a few simple drills, in which each truckman may learn to do thoroughly any portion of the work usually required at a fire, are greatly to be preferred to fancy maneuvers, where each man perfects himself in but one or two duties. After each drill, upon returning to the hose-house, the truck and its equipment should be examined carefully to see that everything is in perfect order. In climbing a ladder the hands should grasp the sides, the body thrown back somewhat, and the feet placed upon the rungs near the sides. As the extension ladders lie upon the truck, the foot at the rear end, the smaller ladder occupies a position *above* the larger. The ladders should be known by their size, and when a ladder is called for the officer should designate it as "36-ft. extension," "26-ft. extension," or whatever it might be.

After the Ladder Company becomes proficient in the more common though necessary maneuvers, other drills, modifications of those here given, may be practiced, the character of the drills

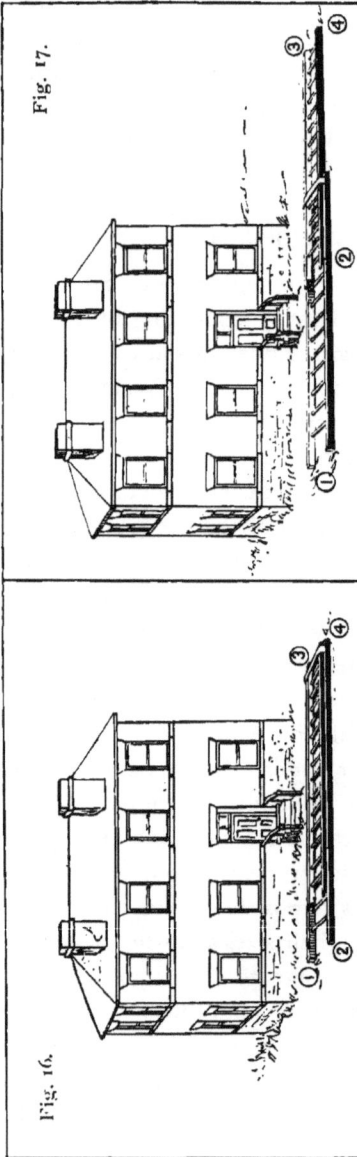

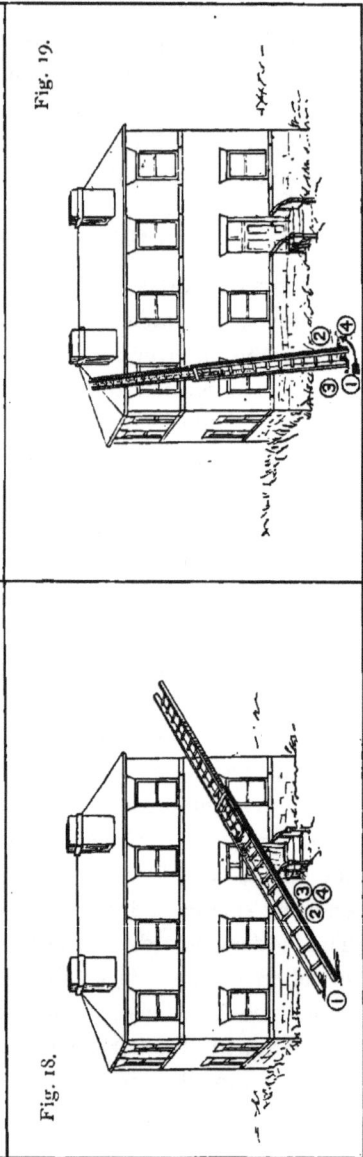

Fig. 17.

Fig. 19.

Fig. 16.

Fig. 18.

depending upon the peculiarities of construction of the various buildings.

RAISING AND LOWERING OF LADDERS.—Ordinarily it will require four men to raise and lower the 36-ft. extension ladder; three for the 26-ft. extension, and three for the 22-ft. single, though if necessary three could raise and lower the first, and two the last named. One man is all that will be necessary to manage the 12-ft. ladder.

Ladders must be *firmly set*, and at a proper distance from the building, neither too near nor too far away.

USE OF AXES, PIKE-POLES, AND CROW-BAR.—Where axes and poles are carried about, great care must be observed in handling them, for by dropping an axe or pole while on the ladder or roof, or by swinging an axe carelessly, a truckman is very liable to injure some one seriously. Axes and poles are very necessary, sometimes, at fires, for exposing hidden fires in walls, ceilings, floors, etc., and although they cannot be used in this capacity at drills, truckmen should become accustomed to carrying them about.

The crow-bar will be useful in forcing off guards, and in prying up timbers, where necessary.

LADDER DRILL NO. I.

Raising and lowering all ladders and climbing to top.

It is the purpose of this drill to make each member of the Ladder Company thoroughly and practically familiar with every detail in the raising and lowering of all the ladders. When the Company is first organized, and from time to time afterwards, this drill should be practiced, slowly and deliberately. It will not be possible to complete the drill at one practice, nor possibly at two.

The truck having been drawn by the company to a point about seventy-five feet in front of some building, and the pin confining the ladders withdrawn, the drill will be conducted as follows:

Thirty-six-foot Extension — First Step.— Four men go to the rear of the truck; Nos. 1 and 2 take the ladder at the foot and pull it from the truck, Nos. 3 and 4 receiving the other end. The ladder is carried to the place designated, and laid upon the ground, about four feet from and parallel to the front of the building, the foot of the ladder resting opposite the point at which the ladder is to be raised. The two ladders which form the extension should occupy the same relative position they had while on the truck — that is, the smaller ladder on top. [Fig. 16.]

Second Step.— No. 1 remains at the foot of the ladder to foot it; Nos. 3 and 4 pull out the smaller ladder until the " dogs" reach the third or fourth rung from the top of the larger ladder, the "dogs" being set by No. 2.   [Fig. 17.]

Third Step.— No. 1 foots the ladder by placing a foot on the upper part of each ladder-foot, heels resting on the ground, hands grasping second rung of ladder, or the sides opposite second rung. Nos. 3 and 4 advance about fifteen feet, take hold of the sides of the ladder, and with one motion raise it above their heads. No. 2 steps in between them to assist, and the ladder is raised to a perpendicular by all three walking forward. [Fig. 18.]

Fourth Step.— The foot of the ladder which is *farthest* from the building is now raised sufficiently to clear the ground, and the ladder is given a quarter turn *toward* the building. [Fig. 19.] The ladder is then allowed to rest in place against the building, and No. 1 will climb to the top.

In lowering the ladder and returning it to the truck the order of procedure is reversed. The ladder is raised to a perpendicular, the same leg as before raised slightly, the ladder given a quarter turn outwards. No. 1 foots the ladder, Nos. 2, 3 and 4 take positions beneath it, and by backing up slowly, lower. When within fifteen feet of the end of the ladder they step from under and the ladder is lowered to the ground. Nos. 3 and 4 raise the "dogs," holding them away while sliding the smaller ladder into place. No. 2 assists in this by pushing the smaller ladder from the end. The ladder is then returned to the truck.

Repeat four times, with the same men in different positions each time, in order that each one of the four may become familiar with each part of the work. Another set of four then go through the drill four times, and this is done until all the members of the company have taken part.

Twenty-six-foot Extension. — With some slight changes in the disposition of the men, the steps for raising and lowering the 26-ft. extension are essentially the same as given for the 36-ft. extension. Leave out No. 2 in the above and there will be no trouble experienced in following the different steps.

Twenty-two-foot Ladder. — This ladder is raised and lowered in the same manner as described above for the others, with the exception that there is no extension.

The reason for raising the ladder leg *farthest* from the building is this: in the event of the firemen losing control of the ladder while in this unstable condition, it would fall against the house, and thus could be righted much more easily than if it fell into the road.

LADDER DRILL NO. II.

Run 150 yards, raise ladders, carry axes and poles to top. " Take up."

At starting, the Captain should be alongside of the men at the head, or a little in advance; there should be two men at the pole, the rest in pairs at the drag-rope. The drill should be made quite slowly at first, the speed to be increased a little each time. The order to "*halt*" should be given by the Captain as the truck arrives opposite the point selected for practice. The two men at the pole pull back, while those at the drag-rope drop it and proceed to take out the ladders as directed by the Captain, who has by this time removed the pin holding the ladders in place. The ladders should be removed in regular order, from top to bottom. The Captain receiving his orders from a Marshal as to where each ladder is to be placed, directs the men accordingly, and the ladders are raised in the manner described in Drill No. I. The men who can be spared first are sent back to the truck for axes and poles, and with these they ascend to the top of the ladders designated by the Captain. If there is an opportunity for using the roof or 12-foot ladder, that also should be sent for.

At the order "take up," all ladders should be lowered and returned to the truck, axes and poles replaced, and the ladders secured by the center pin.

This drill should be repeated three or four times at one practice.

LADDER DRILL NO. III.

Run 100 yards, enter basement with axes, poles and lanterns. "Smudge" fire.

This drill is practised in connection with Chemical Drill No. IV, with a "smudge" fire. After running 100 yards, and the truck is stopped opposite the building, the company is ordered into the basement with axes and poles. The Captain and Lieutenant each take a lantern, which should be lighted either before or after entering the basement. The truckmen should move about carefully, stooping low to avoid as much as possible the inhalation of smoke. The first duty of truckmen is to open windows to allow of the escape of smoke. Those with axes and poles should assist in looking for the fire, and after it is found to use axes and poles as thought best by officers. The twelve-foot ladder could be used to advantage if the windows of the basement are high. After the fire has been extinguished by the Chemical Company, the *debris* should be removed, and the place carefully examined for smouldering embers.

CHEMICAL COMPANY DRILLS.

GENERAL CONSIDERATIONS.—In the following drills of the Chemical Company each member must pay the closest attention to all of the details in the care and management of the extinguisher. It is quite essential that he master each point in connection with the charging, or filling, of the extinguisher, as well as in the discharging of the same. A very large proportion of all fires occurring in hospitals may be extinguished by the proper use of chemical extinguishers, but if improperly used they certainly cannot be expected to do more than other apparatus improperly used.

The extinguishers on the truck, as well as those placed in different parts of the institution, should be frequently examined and kept in good condition. The top should be unscrewed and carefully raised to determine the condition of the bottle and of the solution in the tank. The threads of the top should be lubricated with a very little mineral oil, and the top screwed down tightly. The hose should be tested as to its perviousness by blowing through it while the top is off. The stop-cock should be kept from rusting by the occasional application of a little mineral oil. The hose must be carefully examined for fractures or fissures, which are especially liable to occur near its junction with the tank. If there is any danger of the fluid freezing, owing to severe weather and an exposed position, it would be well to throw into it a handful of common salt. By paying careful attention to these small but important points in the care of the extinguisher, it will last for many years, and be always ready for use. There should always be kept on hand an extra supply of chemical charges, so that there will be no delay in re-charging an extinguisher that has been used. After each run, whether the extinguishers have been handled or not, they must be carefully inspected.

CHARGING AND RE-CHARGING EXTINGUISHERS.— To avoid the possibility of dirt getting into the solution, in filling an extinguisher a fine strainer should be provided, through which the solution could be passed into the tank. The bicarbonate of sodium may be emptied into the strainer, which is held over the mouth of the extinguisher, and water sufficient to fill the tank to within four inches of the top passed through. The bottle of acid, being carefully removed from its saw-dust bed, is first rinsed clear of saw-dust, and then placed in the bottle-holder, which is in connection with the top of the extinguisher, the wheel having first been given a few turns to the left.

Great care is necessary in handling the bottle, as it will break easily, and the sulphuric acid will quickly destroy whatever it comes in contact with. After the base of the bottle is placed in the cup of the holder [Fig. 1, p. 21], the cap which fits over the head of the bottle is raised, the bottle brought under it, and the cap lowered over the head. The spring will hold it in place. The wheel on top is now slowly turned to the right until there is about an eighth of an inch space between the top of the bottle and the firm portion of the cap. The bottle will now be secure, and still loose enough so that any accidental slight turning of the wheel, or any sudden jar the extinguisher might receive, would not cause the bottle to break. The top is now lifted by the two arms, the bottle suspended in the solution, and the top screwed very firmly into place, the threads having been first treated with a little oil, and the washer seen to be in good condition. A piece of easily broken string or twine is now passed through the wheel and about one of the arms, and snugly tied. This is simply to prevent the wheel from being turned either one way or the other without some force being used. The string or twine should not be so stout but that it can be easily broken by a sharp turn of the wheel. The stop-cock being closed, the hose is wound carefully and securely, though not too tightly, about the top of the extinguisher, being held in place by the aid of a quickly detached fastener. A leather strap, fastened about the neck of the extinguisher, with free ends of sufficient length to tie in a bow-knot about the coil of hose, will answer very well.

The stop-cock must be kept closed, and in arranging and securing the hose the nozzle must be *placed upon a little higher level than the solution in the tank*. Attention to these important points will prevent the possibility of an accident that might prove somewhat embarrassing. If the stop-cock is open, and the nozzle hangs lower than the level of the fluid in the tank, there is danger that by siphonage more or less of the fluid will run out, the amount depending upon the position of the nozzle. This of course would not invariably occur, but the danger is great enough to warrant the care and attention suggested.

After the extinguisher has been used, the only thing necessary to do to re-charge it, in addition to the above, would be first to wash out thoroughly the tank, making sure that all of the broken glass has been removed. Before removing the top, however, the stop-cock must be opened, to allow the escape of any gas that may remain. In removing the top the fireman should not hold his head directly over the extinguisher. It is possible that enough gas may remain to lift the top of the extinguisher, with considerable force, several feet. This rarely occurs, but the writer has seen such an accident, the result being a broken nose.

CHEMICAL DRILL NO. I.

Unstrap extinguisher, shoulder, carry, lower, run out hose; wind hose, shoulder extinguisher, carry, place on truck, strap.

This and the drill following, though especially designed for the instruction of new members, should be participated in on each occasion by the whole company. The Chemical Company, together with the Marshals, having assembled at the hose-house, the Chief Marshal or one of the assistants should begin the drill or demonstration by explaining to the company the manner in which the extinguishers are secured to the truck; the positions to be taken by the members of the company in going to a fire; the way in which the extinguishers are released from the truck, and how shouldered and carried. After this is done, the demonstrator should go slowly through the following drill, explaining each step thoroughly. The members of the company in turn repeat the drill.

Facing forward toward the front of the truck, one hand resting on the extinguisher or near it, as in the act of pushing on the way to a fire, the fireman unbuckles the strap that holds the extinguisher in place. Then, lifting it slightly, so that it may clear the rim into which it is set, he tilts the extinguisher forward, the left shoulder being placed a little below its center. Both hands now grasp the extinguisher near the bottom, and by assuming the erect posture the machine is very easily balanced upon the shoulder. One hand (the left) can now steady it in its place, the right being free to open doors, etc. After walking about a moment or two, the extinguisher is placed upon the floor. This is quickly and easily done by grasping the bottom with the right hand, while the extinguisher is allowed to slip forward and downward into the bend of the left arm, and from there is lowered to the floor. After this is done the little strap or string holding the hose in place is unfastened, the hose unwound, the stop-cock opened, and the fireman moves away as far as the hose will allow. (Of course in this drill the breaking of the bottle must be omitted). Reversing now the order of procedure, the stop-cock is closed, the hose carefully coiled, hung about the neck of the extinguisher, fastened there by the small strap or cord, and the extinguisher shouldered. This is accomplished by tilting it towards the left, the left hand supporting it near the top while the right hand seizes the rim at the bottom, and then by a quick lift throwing it upon the left shoulder. In carrying or lifting the extinguisher, the wheel on top must never be taken hold of.

This drill should be practiced until each fireman is perfectly familiar with the different steps.

CHEMICAL DRILL NO. II.

A practical demonstration of the workings of the extinguisher.

For this drill the Company should meet at the hose-house. The first part of the drill should consist of a lecture by the Chief Marshal, or one of the assistants, upon the use of the extinguisher, together with a demonstration of each step taken in charging, discharging and re-charging the same. Upon a table is placed an empty extinguisher, with hose attached; alongside is placed the top of the extinguisher, the box of acid charges and a can of the bi-carbonate of sodium. Near at hand should be pails with water sufficient to fill the tank. The extinguisher is first shown, the manner in which the hose is attached explained, as well as the way in which fluid and gas are forced into the hose. Next the top must receive attention, and every detail of the mechanism of the bottle-holder thoroughly explained. During this demonstration each member of the company should take hold of the articles and examine for himself. The nature of the acid and alkali must be explained, and what the result of a mixture of the acid with a solution of the alkali would be. Attention must be called to the weak point in the acid bottle. After the different parts of the apparatus are shown and are thoroughly understood, the demonstrator should then proceed to "charge" the extinguisher slowly as described above, explaining each step as he proceeds. The extinguisher being now ready for use, it is placed upon the truck, and is ready for the second part of the drill.

A good-sized pile of boxes and barrels having been previously arranged near the hose-house, the demonstrator shoulders an extinguisher and proceeds to this pile with the members of the company following. The pile of boxes and barrels is now lighted, and after it has burned for some time, the manner of turning the wheel and breaking the bottle is shown, the hose being first unreeled and the stop-cock opened.

The demonstrator, while the stream lasts, must call attention to and illustrate how far the stream can be thrown, as well as how close to a fire a fireman may get by reversing his fire-hat; he can also show that the stream may be cut off, by turning the stop-cock, while the fire is being exposed by another fireman with a pole or ax. The fire being extinguished, and the fluid exhausted, the party should return to the hose-house, the extinguisher being carried as in No. 1. The extinguisher should now be cleaned out and recharged as explained above, and the drill is ended.

CHEMICAL DRILL NO. III.

Run 100 yards, shoulder extinguishers, carry into building. " Take up."

This drill may be practiced at the same time that the Ladder Company is having Ladder Drill No. II. Before starting on the 100-yards' run, the members of the Chemical Company should take their positions on either side of the truck, as near the extinguishers as possible, where they can assist in getting the truck along by pushing. When about 50 feet from the termination of the run, the straps securing the extinguishers in place should be unfastened. By the time this is done the truck will be at a standstill, and the four extinguishers are shouldered as in Drill No. I. The captain, securing a lantern, leads the way into the building, to any portion, from basement to attic, as the Chief at the time may order. At each practice a different building should be visited. As it is no boy's play to carry a chemical extinguisher very far, especially up two or three flights of stairs, after running several hundred feet, haste should be made slowly. It is better to *walk* up stairs than it is to *run.* · Arriving at the place designated, the extinguishers are set down, and the hose uncoiled. At the order "take up," the hose is re-coiled, the extinguishers shouldered, the company returns to the truck, and the extinguishers are secured in their proper places. Returning to the hose-house, the extinguishers are carefully examined by the Captain, to ascertain whether or not any of the acid-bottles have been broken during the drill.

CHEMICAL DRILL NO. IV.

Run 100 yards, shoulder extinguishers, carry into building, extinguish "smudge" fire. " Take up."

The material for a "smudge" fire should be gotten together, and preparations made for lighting it before the time set for the drill. The material may be of any substance that will burn slowly, with little blaze, but considerable smoke. A mixture of dampened straw, woolen and cotton rags, with enough dry material to give it a start, answers the purpose very well. A greater or less quantity of this, according to the size of the room and the amount of smoke required, should be placed, not too loosely, in an iron or tin receptacle, the whole being placed in a fire-proof basement room, or other suitable place, where there could be no possible danger of the fire being communicated to any structure. All windows, doors and flues should be closed.

The company having arrived for drill, together with the Ladder Company, a position 100 yards from the building is taken. The Chief should start the fire, remaining in the room until he finds more comfortable breathing near the floor. Then leaving the building, closing the door behind him, he gives the signal, and the companies make the 100-yards' run. The Chemical Company, upon arriving, are ordered into the room where the fire is. The Captain leads the way, the men having shouldered the extinguishers. The fire having been found to be confined to the basement, no ladders are needed, and the laddermen enter the basement, as ordered, with lanterns, axes, poles, etc. [See Ladder Drill No. III.] After entering the room, if it is not known where the fire is located, and on account of the thick smoke this is very likely to be the case, the men with chemicals put them down at some convenient point, crouch near the floor, in order to breathe a purer air and to regain breath, while the Captain, who has a lantern, seeks for the fire. Having found it, he orders: "This way, Chemical Company." The extinguishers are picked up (if the distance is short they need not be shouldered), and carried to a point a short distance from the fire. The extinguishers need never be carried closer to a fire than the length of the hose attached to them. With such a small blaze the Captain will find that one chemical only will be required, and he will direct one of the men to "play away." The hose having been uncoiled and the stop-cock opened, the wheel on top of the extinguisher is turned to the right until the bottle is broken. The last turn sometimes requires considerable force before the bottle will break. While turning the wheel the nozzle of the hose must be pointed toward the fire, and after the stream is on, the fireman managing it must get as close to the fire as he can, lying flat upon the floor if necessary. The stream should be applied as directly upon the burning objects as possible, the stop-cock being used to "shut down" whenever it is necessary that the smouldering substance be turned over. The other Chemical men should remain by their extinguishers until ordered away, so as to be ready to " play away" at any moment. The fire having been extinguished, and everything set to rights, the Chief orders the companies to " take up," and finally to " go home." At the hose-house the empty extinguisher is re-charged, and the others inspected.

This drill should be made frequently, as it accustoms the firemen to working in a thick, smoky atmosphere.

GENERAL DRILLS.

The drills following, or modifications of them, should be taken up as soon as the different companies are fairly familiar with the special drills already described. They should be repeated as often as possible, a general drill once in four weeks not being any too often. It will very readily be seen that without some form of drill whereby the Brigade as a whole might learn to work together harmoniously, as they would be obliged to do at a fire, officers and men would, in the excitement of a fire, become confused, and valuable time would be lost. It is also important that the Marshals have occasional opportunities for studying the manner of directing an attack against fire, as well as to discover and correct mistakes made by individuals and companies. No matter how faultlessly separate drills are done by the different companies, mistakes will happen at a general drill, and the less practice the Brigade has the more serious will be the mistakes made at a fire or at a general practice. A general drill will serve also as a test, not alone of the efficiency of the Brigade, but of the working condition of all the machinery, pumps, water-mains, hydrants, electric alarm, and fire-extinguishing apparatus in general. The drills should be conducted in the same manner as though the Brigade had a *bona fide* fire to deal with.

It is important, while the organization is still young, to have general drills *with previous notification* before having drills without such notice. After a few such drills the Brigade will be prepared to respond to an alarm for a general drill, without previous notification, or for a fire, with little or no confusion or excitement.

It is not thought necessary to go into the details of the general drills, as this has been gone over sufficiently at length in describing the special drills.

GENERAL DRILL NO. I.

Alarm, with notification; day; test of water pressure, etc.

Some time before the alarm is sounded for this drill, a general notification should be given, and just before the alarm is to be sounded firemen should put on their "bunkers." Time should be kept from the moment of the alarm until water is thrown.

At the sounding of the alarm the Chief and First Assistant proceed to the location selected, while the Second Assistant and the members of the different companies go to the hose-house. Here the companies take their proper positions, the Captain or Lieutenants of hose companies securing carts and calling, "This way, No. ——," giving the number of the company of which he is an officer. The Assistant Marshal calls out plainly the location, and the companies as fast as they are formed leave the house. As already pointed out, it will often be necessary to mix companies to some extent, on a general turnout. In order to do this to the best advantage, and with the least amount of confusion, the Marshal at the hose-house must act quickly and with good judgment. No rule can be laid down as to how this should be done. After practicing this general drill a few times, the manner in which companies are best formed will become apparent.

After starting the majority of the companies, the Assistant Marshal should run ahead to assist as much as possible in placing the different companies, taking his orders from the Chief or First Assistant.

Each company, upon arrival, must be put to doing something *at once*, and be kept busy until the drill is over. The first hose company ready should be ordered to "play away." The disposition of the various companies must at each practice of this drill be determined by the Chief. He should have a definite plan of action based upon an imaginary fire. Combinations and modifications of the various special drills can be worked out; points not mentioned in any of the special drills may be practiced, as for instance : the company throwing water is supposed to have burst a length of hose; the order is given to "shut down," and the burst length removed and either another length supplied, or the separated lengths brought together and connected. At the same time this is being done another hose company is called to the place occupied by the other, and ordered to "play away." Another useful maneuver would be the pulling up to a high window or roof of a lead of hose by the use of the rope carried on the ladder truck for this purpose. Companies should be changed about during the drill, that they may become acquainted with the various orders. When the drill is finished the order to "take up" must be given by the Chief to the Captains of the companies separately, no company to "take up" until so ordered by the Chief. The Captains should give the orders respecting the details. After "taking up," the companies should remain on hand until ordered by the Chief to "go home." Arriving at the hose-house, and the apparatus having been looked over carefully, the companies are all dismissed with the exception of the companies having used water. They should remain to change the wet hose upon the cart for dry.

GENERAL DRILL NO. II.

Alarm without notification; day; imaginary fire in attic, basement or other place.

The location of the imaginary fire having been decided upon, an alarm, *without previous notification*, is turned in. The drill is to be conducted as though there really was a fire, except that no water is to be used in-doors. Where fire pressure is used, an outside stream, thrown on the roof, for instance, should be ordered. If it is not desirable to use water, word can be sent to the Engineer to take off the extra pressure. Each time this drill is given a new location should be selected. In this way the officers may make a practical study of the manner in which a fire in different parts of the various buildings may be fought. After the Brigade is quite familiar with the special drills, this general drill should be frequently repeated.

GENERAL DRILL NO. III.

Alarm, without notification; day; "smudge" fire in basement.

A "smudge" fire should be made in a basement, as for Chemical Drill No. IV. The alarm is then turned in, without previous notification. Upon the arrival of the Brigade the Chemical and Ladder Companies are ordered into the basement, as at special drills. One hose company should be ordered into the basement, another into the building just over the fire, and another to take position at one of the basement windows.

GENERAL DRILL NO. IV.

Alarm, without notification; night; false alarm.

This drill needs no further explanation than this: that the alarm should be turned in just after all have retired, say between ten o'clock and midnight, or it may be better to do so just before the rising hour. This drill should be given once in six months, unless there has been a night alarm other than

the drill during that time.   Firemen should not forget their
" bunkers."

EXHIBITION DRILLS AND CONTESTS.

Exhibition drills must necessarily differ in some respects
from the ordinary fire practice, for the reason that much of
the work done by the Brigade in a regular drill would be not
only out of sight of the spectators, but if seen would not be
fully appreciated.   Something more showy is necessary, and
while not as useful to the firemen as would be a regular drill,
the exhibition encourages them to put forth their best efforts.
A few hints as to these exhibition drills will suffice.

The plans for the exhibition should be thoroughly made,
and they should be so well understood by the Marshals as to
preclude the possibility of failure.   If it is simply the desire to
show how quickly the Brigade can respond to an alarm without previous notification, no planning will be necessary, except
to provide for a *full* turnout, something that could not be done
in case of fire, except at night.   It is not *essential* in any case
that the whole Brigade be previously notified, but it is often
desirable.   It *is* necessary, however, in order that a creditable
exhibition may be given, that certain preparations be made in
advance, such as : an understanding between the Marshals as
to the form of drill ; time and place ; the placing of spectators,
etc.   Have all the work done, as far as it is possible, in sight
of the spectators.   In building a pile for a fire for the Chemical
Company to extinguish, do not make it so large that it will
require more than two extinguishers to put it out, nor yet so
small that it will be quenched in a moment.   In making such
a pile use dry, quickly-burning material, loosely put together.
An exhibition drill could be made up of some of the regular
drills and contests.   Some such arrangement as the following
may be taken as an illustration :

EXHIBITION DRILLS.

A favorable spot having been chosen, a pile for a bonfire thrown up, and the spectators settled, the alarm is sounded.* As the Brigade arrives the hose companies are directed to the proper hydrants by the Assistant Marshals, and all the hose is laid and pipes attached, the leads converging to a point indicated by the Chief. The first company ready is ordered to "play away," and then the second. The others must keep their hose dry. The order to "play away" can be given, in this instance, just before the pipe is attached. The Ladder Company, upon arrival, is ordered to place its ladders against some adjoining building, the Chemical Company removing their extinguishers to a short distance. The order "shut down" is soon given to the two hose companies using water, followed by the order, "take up," to all the companies. The Brigade should then pass in review before the spectators, after which may be introduced, in the order named: Trial of Speed by Ladder Company; Hose Contest No. II. by the two companies having the best records; Hose Drill No. V. by the remaining two companies; lighting of bonfire, with work by Chemical Company.

The following Hose Company contests and Ladder Company trials of speed will aid materially in keeping up interest in the work of drilling, which might become irksome without some such incentive.

HOSE CONTEST NO. I.

Run 100 yards to hydrant, and connect; lay 250 feet of hose; attach pipe.

The cart and hydrant having been thoroughly inspected by the Chief, the company takes its position 100 yards from the

---

* With such preparations it would be impossible to keep the boys from getting an idea of what was going to happen next, so that to see them rushing towards the hose-house almost before the alarm sounds must be expected.

hydrant. The hydrant-cap must be on by three turns, and the connections at pipe and hydrant must be made by not less than three turns. The reel may be freed from fastenings. The start is made by a signal from the starter, and the finish announced by the chief calling "*Time.*" "Time" is not to be called until the pipeman has made his coupling and dropped the pipe to the ground, and the hydrant-man has made the hydrant connection, fitted the wrench to the rod, and stepped away. The connections must be examined and found to be correct before the time made is recorded. [See Hose Drill No. I.]

HOSE CONTEST NO. II.

Run 100 yards to hydrant, connect, lay 250 feet of hose, attach pipe; "play away," water at fire pressure.

This contest is the same in every particular as No. I., except that water is used. "Time" is called when water comes from the pipe (which is held by the pipeman) and the hydrant gate is wide open. The order to "play away" may be given, in this instance, as the cart passes the hydrant, for generally by the time the hydrant-man is ready to turn on water it will be safe for him to do so. He should, however, turn slowly or rapidly, according to how nearly ready the pipeman is. In order that no time be lost in waiting for water after everything else is ready, the hydrant-man should endeavor to have water burst from the nozzle just as the pipe is in position. With a little practice and well-understood signals between the Captain and hydrant-man, this contest drill can be accomplished almost as quickly as No. I.

HOSE CONTEST NO. III.

Run 100 yards to hydrant; connect; lay 250 feet of hose; attach pipe; break couplings; make couplings.

[See Hose Drill No. II.] Reel may be free from fastenings. All connections must be made by at least three turns.

The hydrant-man must make the first connection at hydrant, and the pipeman at pipe as in Contest No. I.   Two of the company then break and make all connections as in Hose Drill No. II.   "Time" is called when the pipe is turned into place the second time, and is dropped to the ground.

LADDER COMPANY TRIAL OF SPEED NO. I.

Run 100 yards, raise three ladders, climber to top of each.

This is very similar to Ladder Drill No. II., except that no axes or poles are to be used.   A certain place for footing each ladder should be previously decided upon.   In each case the man who foots the ladder should mount to the top.   When the last one to touch the top round of his ladder has done so, "time" should be called.   By careful practice this trial of speed may be done very smoothly, and in remarkably quick time.   The three ladders can be taken from the truck almost simultaneously, so that the three climbers will reach the top of their respective ladders at about the same time.   This and the trial following would be very effective in exhibition work.

LADDER COMPANY TRIAL OF SPEED NO. II.

Run 100 yards, raise 22-foot ladder to perpendicular, climber to top.

The four best ladder raisers and the best climber should be selected, and so placed at the sides of the truck that they can drop back without trouble.   A line should be marked across the road or lawn 100 yards from the starting point, across which line the truck will pass without slackening speed. When within about 75 feet of the line, the selected men should fall behind and remove the 22-foot ladder from the truck.   The climber stops at a point 20 feet back of the line.   The foot of the ladder is placed on the line indicated, and the ladder is raised almost to a perpendicular.   The climber runs from his position, jumps over the stooping truckman footing the ladder, and climbs to the top, the four men steadying the ladder as he climbs.   After some practice this can be done very quickly.

CHAPTER VII.

THE FIGHTING OF FIRE.

IN the fighting of fire, while general principles are applicable to most cases, no special mode of treatment or action can be laid down for any single instance. As each fire differs in many particulars from all other fires, so the treatment will differ in each case. The indications must be met as they arise. The *particular* treatment necessary in any case must be decided upon at the time of its existence; it can not be done before. The plan of action adopted at each fire will depend upon many things — the time of day, the condition of the atmosphere; the direction and velocity of the wind; the construction and uses of the building; the location of the fire; its extent; manner and rapidity of its progress, etc., as well as upon the kind and reliability of apparatus and the efficiency of the Brigade and the judgment of its officers.

OBEDIENCE TO ORDERS.

One of the most essential things for firemen to learn is the necessity, at a fire, for the rapid execution of the plans of the officers in charge. That this may be done, *each order*, *when given by the proper officer*, no matter how unnecessary it may seem to be to the one receiving it, *must be promptly obeyed*. A single moment's delay might be the means of causing irremediable damage. There should be no questioning, no stopping for explanations once an order is given by the proper person. This is as necessary for the protection of the firemen themselves as it is for the protection of property and of the lives of others. A company may hesitate just long enough, after receiving the order to "back out," for instance, to have all avenues of escape cut off, or to get caught under a falling wall.

Suggestions and orders from outsiders, those who are simply lookers-on, are frequently given at a fire. Firemen should never pay any attention to such, for the danger of so doing would be as great, almost, as the disobeying of orders from the proper source.

COOLNESS AND PRESENCE OF MIND.

At a fire, officers and men should endeavor to work with as little excitement as they would in going through an ordinary drill. A single excited individual, it matters not what his position may be, is liable to interfere seriously with the work of the rest of the Brigade. Cool-headedness should be especially cultivated by the officers; they should strive to be self-possessed under the most trying circumstances, and equal to any emergency. A cool, clear head is requisite for the solution of unexpected difficulties, which are constantly arising. The members of the different companies will generally emulate the example of the company officers, and these in turn that of their superior officers. There should be an absence of all unnecessary noise and running about. If the Brigade is well drilled, but few orders will be required, and there need be little if any loud talking except in the giving of orders. Firemen should do what they are told to do, calmly and deliberately, relying upon the judgment of their officers.

GIVING THE ALARM.

There should be the least possible loss of time between the discovery of fire and the turning in of the alarm, as each moment of time is of the greatest importance to firemen. Each second's delay adds to the difficulty of extinguishing fire. Not only should employes be instructed as to the *manner* in which an alarm is to be given, but the fact should be thoroughly impressed upon their minds that there must never be any delay, after a fire is discovered, in giving the alarm. If the fire be very small, and could of a surety be *easily* extinguished with the means at hand, then it would not be necessary to call out the Brigade; but it would in any case be better to call out the firemen needlessly than to wait until the fire has made headway before giving the alarm.

RESPONDING TO ALARM.

Immediately upon an alarm of fire each member of the Brigade must act, and act promptly. To go to the hose-house, and from there to the fire with the apparatus as quickly and as orderly as possible must be his first thought. There should be, however, no wasting of strength or breath by rapid running. If this is done the firemen arrive at the fire panting for breath, and totally unfit to enter a hot or smoky atmosphere. If firemen will always think to have their "bunkers" ready by their beds, there is no reason why even better time could not be made at night than during the day.

AT THE HOSE-HOUSE.

If the hose-house be centrally located, there need be but little delay in the starting of apparatus. The ladder truck should be gotten out as soon as there are enough members of the Ladder and Chemical companies to start it. The rest of the members of the two companies could join in afterwards, the bell on the truck acting as a guide, especially at night. The first Hose-Company Captain or Lieutenant to arrive should secure a cart nearest the door and call out, "This way, No. —" naming the number of his company. A hose-cart could be started out if provided with a captain or lieutenant, a lineman, and a hydrant-man. It will not always be possible, nor will it be necessary, for hose companies to leave the house without more or less' mixing. As will be seen by reference to the chapter on "Organization," it should be the duty of one of the Assistant Marshals to go to the hose-house upon an alarm, there to assist in forming the companies and getting them started, giving the location, and directing the later arrivals what to do. After practicing some of the general drills a few times the question of what is best to be done at the hose-house will become easy of settlement, if properly studied by the officers.

SMOKE AND HEAT.

One of the greatest obstacles a fireman has to contend with at a fire, especially in its early stage, is the presence of smoke. It often prevents a near approach to an otherwise easily accessible fire, and in consequence much valuable time is lost. The quantity of smoke coming from an open door or window, however, is not always an indication of the amount to be found within, for it is often the case that firemen may breathe comfortably in rooms from which are pouring, by window or doorway, great volumes of smoke. By stooping low it will generally be found that an entrance to the building or room may be safely made. The condition of the atmosphere of the interior may then be observed. This fact should be impressed upon the mind of every fireman — that heat and smoke, by their lightness, tend to rise; that therefore the intensity of heat and the density of smoke increase rapidly as the ceiling is approached, and conversely, the heat becomes less oppressive and the air much freer from smoke as the floor is neared. By crawling upon the hands and knees, or by lying flat upon the floor, firemen will find they can get about, or remain in a very smoky room for a considerable length of time, where the upright position could not be maintained for more than a moment or so.

If the amount of smoke and heat in any way interferes with the firemen's efforts in finding or extinguishing the fire, there should be no delay in providing for the rapid escape of both heat and smoke. Windows should be lowered *from the top;* if this cannot be done, then at least one should be broken at the top, the others being raised from the bottom. Of course if raising the windows will accomplish the object, then there is no necessity for breaking one, but one window open at the top will allow of the escape of more heat and smoke than six raised from the bottom. It becomes necessary at times, on account of the great amount of smoke and heat, for an opening to be made in the roof, through which the smoke and heat may escape. The objection will be raised to these suggestions that the opening of windows and doors, and cutting holes in the roof, will create a draft, and thus favor the rapid spread of fire. This is partly true; if the fire is so situated (in an unoccupied building, in a closet or single apartment, for example) that the escape of the inmates will not be prevented by the smoke, then windows and doors should be kept closed until the firemen are ready to attack the fire. But for firemen to enter a building or apartment already filled with thick, blinding smoke and a furnace-like atmosphere, and then to refrain from opening windows and doors, for fear of creating a draft, would be utter folly. The firemen would be fighting in the dark, guessing at the location of the fire, wasting time, and endangering their lives needlessly.

When ample provision is not made for the escape of heat and smoke, it often happens that there is a sudden reversal of air currents, the scorching air above changing places with the cooler atmosphere nearer the floor. This might prove very serious to anyone so unfortunate as to be in the room at the time. Again, smoke itself, when dense and sufficiently heated, is highly combustible.

A wet handkerchief, or cloth of any kind, held over the mouth and nose, makes an excellent respirator. For hosemen, the use of the spray nozzle attachment to the play pipe is a great protection against excessive heat and smoke.

Reversing the fire-hat, so that the long peak comes in front, will give additional protection against heat.

RETREAT.

The officers of the Brigade should have in mind, during the progress of a fire, the many dangers by which firemen are surrounded. Each possible source of danger should be carefully studied, and in each instance the officer should have in mind the way in which retreat could be safely made in case of necessity. The Chief should keep a sharp look-out for tottering walls, sinking floors, and weakening roofs. He should not wait until the last moment before giving the order to "back out" or "back down," when he apprehends danger.

INCIPIENT FIRES.

Fires involving the loss of life and the destruction of much property are frequently the result of either the neglect or improper management of incipient fires. In places provided with the proper facilities for fighting fire, a well-drilled Brigade, and with a method, almost instantaneous in its action, of notifying the firemen of the presence and whereabouts of a fire, there is no reason why so many incipient fires should be allowed to become destructive conflagrations. The proper management of these fires by the Brigade will depend greatly upon the good judgment used in each case by the officers in charge as well as upon the amount of attention the drilling of the Brigade has received.

The few observations following may be of service to the Brigade in the treatment of fire in its early stages:

— Pending the arrival of the Brigade in case of the smaller

fires, efforts should be made by those in the vicinity of the fire either to extinguish it, or else endeavor to hold it in check with the means at hand. The safety of the inmates is to be first considered, however.

— If a fire in a closet, wardrobe, or other small apartment cannot be reached with water from pails, the doors and windows of said apartment should be tightly closed until after the arrival of the Brigade.

— In using water from fire-pails, it should be judiciously applied, not thrown haphazard. If possible, it should be thrown on to the bed of the fire rapidly and continuously, but in comparatively small quantities at a time. The use of a cup or dipper, or better still, of a small hand force-pump, if one is provided with the pails, for distributing the water where it will do most good, would be a much better plan than to throw the whole pailful at once. Especially if the fire be situated out of easy reach will it be folly to try to extinguish it by attempting to empty the pail at once.

— *A woman should never attempt to extinguish a fire by trampling upon it, no matter how small it may be.* A spark, flying upward, is very liable to lodge in some portion of her wearing apparel, soon to light up a fire that may cost her her life.

— Rugs, carpets, and blankets are generally convenient, and if spread over small fires occurring on tables, beds, or upon the floor (such as burning table-spreads, bed clothes, piles of wearing apparel, etc.), will often extinguish the fire, or at least hold it in check or prevent its spread, by depriving it of air. The efficiency of this method will be greatly increased by throwing water upon whatever is used for smothering the fire.

—No matter how small the fire, nor how quickly it has been extinguished, the Chief should not feel free to send the firemen away until he has proved beyond a doubt that no smouldering ember has been overlooked. Burned or charred lathing, flooring, etc., must be wholly removed, or partially removed, so that a thorough inspection may be made of that which would otherwise be hidden from view. After fires in clothes-closets, store-rooms, bedding, and the like, the partially burned or charred material should be spread out and carefully examined for sparks or smouldering fragments.

—In combating fires in walls, floors or ceilings, the attention of the firemen should not be alone directed to its point of origin, for while they are congratulating themselves upon an easy victory, the fire may be making rapid progress upward, between the studding, to attic or roof. Each concealed space in wall or floor, unless provided with fire-stops, becomes a flue, through which fire will travel with remarkable rapidity. Bearing this in mind, firemen must not rest content until they are sure they have extinguished *all* of the fire. The upper opening of the flue should be watched during the progress of the incipient fire below, an extinguisher or lead of hose being near at hand and ready for instant use.

—It is sometimes necessary to tear down plastering, lathing, base-boards, partitions, door and window-casings, etc., and to tear up flooring, that the hidden portion of a fire may be reached and extinguished. Like all necessary things, this is as often overdone as it is underdone. The former is to be preferred to the latter. A well-disciplined, well-drilled and well-officered brigade will refrain from either extreme, but will do as nearly just what is essential as the circumstances will allow,— no more, no less.

—If a fire is known to exist, and yet is concealed from view, it must be diligently searched for, and when located, *thoroughly* exposed, so that it may be acted upon to the best advantage by pail, extinguisher or hose. If the officers of the Brigade know, by previous observation and study, the peculiarities of construction of the various buildings, and have paid attention to some of the causes of fires, they will have but little trouble in discovering the location of concealed fires, and in properly exposing the same.

— The danger from smoke and the rapid spread of fire is much greater from those fires having their origin in the basement or lower stories, than from those originating in the roof, attic or upper stories.

SMOULDERING FIRES.

In the spontaneous ignition of oiled rags, and in the slow combustion of woodwork, etc., not exposed to the air, it frequently happens that the material will smoulder and smoke for many hours, sometimes for several days, before bursting into active flame, owing to the absence of a sufficient amount of air to produce active combustion. The presence of smoke or the odor of anything burning in a building should never be neglected, and a most thorough search begun at once and not given up until the cause for the same has been found. In very mild, uncertain cases the notifying of the Chief or other officer would be sufficient, but if the odor be distinct, and the cause be not immediately apparent, then there should be no hesitancy whatever in turning in an alarm. Especially should there be no delay at night. Every nook and corner in every part of the building, from cellar to attic, should be thoroughly searched. These fires may be caused by oiled rags in closets,

chutes, between floors; by heat-conductors in too close proximity to wood-work; defective furnaces; dirt in registers, etc.

FIRES IN COAL PILES.

In endeavoring to extinguish fires occurring in the interior
of piles of bituminous coal, it is useless to throw water upon
the outer surface.   The small amount that finds its way through
the coal to the fire, instead of extinguishing, adds to the fire.
Water must be made to reach the burning coals in as direct a
manner and in as large quantities as possible.   The use of
three-inch iron pipes, driven down into the burning mass,
through which water may be forced from the hose, has been
found to be the most successful means employed.   In the absence of iron pipes, the digging of one or more holes down to
the burning coal, into which streams of water are turned, will
generally bring about the desired result.

SPECIAL SUGGESTIONS.

HOSE COMPANIES.

Do not fight fire by attacking flame and smoke.

It is quite useless and a great waste of time to direct a
stream of water against flame or into smoke.   The presence of
a great amount of suffocating smoke does not necessarily indi-
cate that the fire is an extensive one.   A small piece of smouldering wood, or a few handsful of burning straw or woolen rags
are often sufficient to fill a house so full of smoke as to make
it very uncomfortable for people to remain in it.   Nor does
the amount of smoke alone show the firemen the exact location of the fire, so that it would be useless for him to direct his
stream at smoke, no matter how dense it might be.   A stream
of water should not be thrown into a shaft or chute of any kind
simply for the reason that smoke is issuing from it.

Throwing a stream of water into the *flames* with the object

in view of extinguishing the fire, is nearly if not quite as unsatisfactory as trying to put out the fire by wetting down the
smoke. It is a temptation to a fireman, whenever a blaze of fire
shows itself, to "hit it," but while his stream of water is either
being converted into steam, or is passing through the flame to
fall in a place where it is not needed or will do actual harm, the
fire itself, at its base, is making steady progress. In this way
many a small, incipient fire is allowed to become a serious one.

Water, to be effective, must be applied directly to the burning coals.

This fact must certainly be evident to all, yet it is not in-
frequently disregarded by firemen in their efforts to extinguish a fire. Pains should be taken to discover and expose, as
much as possible, the burning objects,—*the bed of the fire,*—and
the stream of water thrown directly, and from the nearest
practicable point, *against the burning coals.* The more solid the
stream is when it reaches the burning mass the more effective
will it be. A very feeble stream of water, reaching a fierce
fire in the form of spray, really adds to its fury by supplying
fuel, for water, under certain conditions, is combustible. It is
not always possible to expose the bed of the fire, nor to apply
the stream directly to it. In this case the stream must be
directed against the surrounding wood-work. In approaching
a good-sized fire for the purpose of throwing water upon the
burning coals, it should be done from the rear of the fire, if
convenient; that is, where there is the least heat and smoke.
A much nearer position can be thus taken by the firemen, and
the efficiency of the stream is not in danger of being lessened
by passing through flame before reaching the coals.

---

*Keep the adjoining structures wet, that spreading of the fire may be hindered
or prevented.*

Next in importance to fighting fire where it exists is *the prevention of its spread,* especially when it is gaining headway. This is done, in a measure, by thoroughly wetting and keeping wet those structures immediately surrounding the burning objects, as well as those which are in danger of being reached by the flames or by flying embers. With a wind blowing, and the fire spreading, it is quite as essential to have a stream of water anticipating the advance of the flames as it is to fight the body of the fire. The fire can be best headed off by having the stream of water thrown from a position at one side or in front of the spreading flames, so that in reaching the exposed structures the water would not first pass through fire.

———

The order to "play away" should not be given until the pipeman is in position to play on the fire.

For the following reasons : 1.— Empty hose is much easier to handle, and therefore can be gotten into position more quickly than hose filled with water. 2.— A heavy force of water coming to the play-pipe unexpectedly, may find the pipeman from his position on ladder, stairs, roof or window-sill, totally unprepared for the shock, and cause him, and probably others, to fall. 3.— The stream is liable to strike almost anywhere except where it is needed, and possibly cause much damage and create needless confusion. 4.— Time is lost, not saved, the fire, in the meantime, rapidly gaining headway.

CHEMICAL COMPANY.

Do not unreel hose until the proper position is reached for playing upon the fire.

If this precaution be not observed the hose is liable to become torn or otherwise injured by dragging upon the ground or floor. It is also liable to trip up the carrier of the extinguisher.

Do not charge the Chemical until ready to play upon the fire.

In other words, do not waste the fluid by throwing it where it will do no good.

Get as close to the fire as possible, and direct the stream upon the burning coals.

The shorter the range, the more effective will be the stream.

An extinguisher should never be strapped to the back, or in any way fastened to any portion of the body.

It is just as easy to carry the extinguisher balanced upon the shoulder, and not nearly as dangerous.

LIFE-SAVING CORPS.

Upon an alarm of fire the members of the Life-Saving Corps must proceed directly to the location given. Their special duties will vary greatly, from simply reassuring patients and preventing and subduing excitement, to removing large numbers of patients from a burning building to a place of safety, and rescuing and caring for those who are sick, insensible or otherwise helpless.

Each room should be carefully searched for patients — in closets and corners ; behind doors ; in beds and under them. When it is certain that the room contains no one, the door should be closed and locked (unless there be good reasons why this should not be done), to prevent the return of patients into it.

After patients are removed from the building they should be kept together and taken as soon as possible to a safe place.

Patients should be prevented from returning to the burning building by placing at each exit-door a member of the Corps or one of the ward attendants.

In rescuing an insensible or otherwise helpless person, who is too heavy to be easily carried, it could in most instances be easily done by dragging the bed, upon which the patient is lying, along the floor. If stairs are to be descended, a bed with quite a heavy person upon it may be guided in its slide by two persons,— one at either end.

A person whose clothing is on fire should be prevented from running about, and if possible made to lie down. In approaching such a person, the rescuer must hold in front of himself a rug, blanket, coat, or whatever he has been able to secure for the purpose. This he should wrap about the other, the intention being to smother the fire by excluding air. If the wearer of the burning garments is standing, he must be made to lie down, and then rolled snugly into the article used for smothering the fire.

www.ingramcontent.com/pod-product-compliance
Lightning Source LLC
Chambersburg PA
CBHW030626270326
41927CB00007B/1330